T0275971

SpringerBriefs in Electrical and Computer Engineering

More information about this series at http://www.springer.com/series/10059

Christoph Guger · Gernot Müller-Putz
Brendan Allison
Editors

Brain-Computer Interface Research

A State-of-the-Art Summary 4

 Springer

Editors
Christoph Guger
g.tec Guger Technologies OG
Schiedlberg
Austria

Gernot Müller-Putz
Graz
Austria

Brendan Allison
Department of Cognitive Science
UC San Diego
La Jolla, CA
USA

ISSN 2191-8112   ISSN 2191-8120   (electronic)
SpringerBriefs in Electrical and Computer Engineering
ISBN 978-3-319-25188-2   ISBN 978-3-319-25190-5   (eBook)
DOI 10.1007/978-3-319-25190-5

Library of Congress Control Number: 2015953778

Springer Cham Heidelberg New York Dordrecht London

Printed on acid-free paper

Springer International Publishing AG Switzerland is part of Springer Science+Business Media
(www.springer.com)

Contents

Brain-Computer Interface Research: A State-of-the-Art Summary 4

Christoph Guger, Brendan Z. Allison and Gernot R. Müller-Putz

What Is a BCI?

This book presents the latest research in brain-computer interface (BCI) systems. A BCI is a device that reads voluntary changes in brain activity, then translates these signals into a message or command in real-time. Early BCI work focused on providing communication for persons with severe movement disabilities. These patients have little or no ability to communicate in other ways, such as by speech, typing, or even most assistive technology (AT) systems meant for disabled users. For these patients, BCIs can provide the only means of communication.

All BCIs involve four components:

(1) Sensors measure brain activity.
(2) Automated software converts this activity into messages or commands.
(3) This information affects an output device in real-time.
(4) An operating environment controls the interaction between these components and the user.

Most BCIs rely on the electroencephalogram (EEG). These signals (also called "brainwaves") can be detected with electrodes on the surface of the head. Thus, these "noninvasive" sensors can detect brain activity with very little preparation.

C. Guger (✉)
g.tec medical engineering GmbH, Guger Technologies OG, g.tec medical engineering Spain SL, g.tec neurotechnology USA, Inc., Schiedlberg, Austria
e-mail: guger@gtec.at

B.Z. Allison
Cognitive Science Department, University of California, San Diego, CA, USA
e-mail: ballison@cogsci.ucsd.edu

G.R. Müller-Putz
Laboratory of Brain-Computer Interfaces, Graz University of Technology, Graz, Austria
e-mail: gernot.mueller@tugraz.at

© The Author(s) 2015
C. Guger et al. (eds.), *Brain-Computer Interface Research*,
SpringerBriefs in Electrical and Computer Engineering,
DOI 10.1007/978-3-319-25190-5_1

1

Some BCIs are "invasive", meaning that they require neurosurgery to implant sensors. These BCIs can provide a much more detailed picture of brain activity, which can facilitate prosthetic applications or surgery for epilepsy and tumor removal.

Automated software for BCIs requires several different stages. The raw data from the brain needs to be amplified and filtered. Spatial filtering can use signals from different electrodes to further reduce noise and improve signal quality. Additional preprocessing might further reduce noise, eliminate unnecessary data, or make data easier to process. Next, software must identify meaningful activity. This challenge depends on the BCI. For example, if the goal of the BCI is to detect hand movements to control a prosthetic device, then the software must determine which hand movement(s) the user is imagining.

This information is then sent to an output device. Common examples include a monitor, Smart Home system, prosthetic or other robotic device. The monitor might present feedback, such as whether the user is imagining left or right hand movement, which could move a cursor. Smart Home controls can control lights, doors, curtains, and other systems. Prosthetic limbs could restore control for many patients, and robotic devices such as a Functional Electrical Stimulator (FES) or robotic exoskeleton could be used for therapy.

Finally, an operating environment must provide seamless, real-time communication between the different software modules and the user. This can be especially challenging in some BCIs that have to work with EEGs, other input signals, and complex robotic devices. Some open-source software platforms are designed to help reduce the burden on academic and clinical researchers developing a BCI operating environment.

In the last several years, BCIs have improved in many ways. Dry electrodes can measure EEG with much less preparation time. New "passive" BCIs focus on monitoring changes in alertness, emotion, or other factors that do not distract the user. Different research groups have explored BCI technology to help communicate with persons who are (mis)diagnosed as comatose. Other work extends BCIs to help persons with stroke, memory or attention deficits, or other conditions. For example, BCIs that are integrated with a functional electrical stimulator (FES) or robotic exoskeleton can help people train to improve motor function that was lost due to stroke. Many of the chapters in this book present some of the top groups' newest work with BCIs to help new patient groups.

The Annual BCI-Research Award

G.TEC is a leading provider of BCI research equipment headquartered in Austria. Because of the growth of BCI research worldwide, G.TEC decided to create an Annual BCI-Research Award to recognize new achievements. The competition is open to any BCI group worldwide. There is no limitation or special consideration for the type of hardware and software used in the submission. The first Award was presented in 2010, and followed the same process that has been used in subsequent years:

- G.TEC selects a Chairperson of the Jury from a well-known BCI research institute.
- This Chairperson forms a jury of top BCI researchers who can judge the Award submissions.
- G.TEC publishes information about the BCI Award for that year, including submission instructions, scoring criteria, and a deadline.
- The jury reviews the submissions and scores each one across several criteria. The jury then determines ten nominees and one winner.
- The nominees are announced online, and invited to a Gala Award Ceremony that is attached to a major conference (such as an International BCI Meeting or Conference).
- At this Gala Award Ceremony, the ten nominees each receive a certificate, and the winner is announced. The winner earns $3000 USD and the prestigious trophy.

The scoring criteria have also remained consistent across different years. These are the criteria that each jury uses to score the submissions. Given the intensity of the competition, nominated projects typically score high on several of these criteria:

- Does the project include a novel application of the BCI?
- Is there any new methodological approach used compared to earlier projects?
- Is there any new benefit for potential users of a BCI?
- Is there any improvement in terms of speed of the system (e.g. bit/min)?
- Is there any improvement in terms of accuracy of the system?
- Does the project include any results obtained from real patients or other potential users?
- Is the used approach working online/in real-time?
- Is there any improvement in terms of usability?
- Does the project include any novel hardware or software developments?

In 2014, G.TEC introduced second and third place winners. Because of increased interest in the Annual Awards, and increasing number of high-quality submissions, it seemed appropriate to honor second and third place as well. The jury selected these project based on their total scores. This change was well-received by the nominees and the BCI community, and will continue in later years.

Thus, this year's jury had an especially difficult task. Scoring was even more important, as small changes could influence second and third place. Indeed, the winners (presented in the concluding chapter) scored very close to other nominees this year. However, the jury spent a lot of time reviewing the 69 submissions, and the three winners were carefully chosen. The 2014 jury was:

Gernot R. Müller-Putz (chair of the jury 2014),
Deniz Erdogmus,
Peter Brunner,
Tomasz M. Rutkowski,
Mikhail A. Lebedev,
Philip N. Sabes (winner 2013)

Consistent with tradition, the jury included the winner from the preceding year (Prof. Sabes). The chair of the jury, Prof. Müller-Putz, is a top figure in BCI research and leads the prestigious BCI lab in Graz, Austria. Prof. Müller-Putz said: "I was very fortunate to work with the 2014 jury. All of the jury members that I approached chose to join the jury, and we had an outstanding team."

The Gala Award Ceremony has become a major annual event. These events attract several dozen BCI researchers each year, and major BCI conferences are scheduled around this evening dinner event to ensure that attendees can make it. The ceremony occurs at a formal location near the conference center, and includes beverages and several courses of gourmet dining. The annual announcement of the winner has always been suspenseful, and the addition of second and third place winners has made the ceremonies even more exciting.

In 2014, the Gala Award Ceremony was hosted at the prestigious Hotel Das Weitzer. This location had excellent dining and presentation facilities, and was close to the Sixth Annual BCI Conference in Graz, Austria in September 2014. The chair of the jury and the other two book editors introduced the Award, reviewed the ten nominated projects, presented certificates to the nominees, and then announced the winners.

Fig. 1 This picture shows the Gala Award Ceremony in Graz

Fig. 2 This image shows a group of six BCI researchers from different countries discussing their work at the Gala Award Ceremony. Facing the camera, from *left* to *right* David Ryan and Samantha Sprague (USA). On the other side of the table, from *left* to *right* Gernot Mueller-Putz, Chairman of the 2014 Jury (Austria), Andrea Kuebler (Germany), Donatella Mattia (Italy), and Nick Ramsey (Netherlands)

Fig. 3 One of the nominees, Dr. Jeff Ojemann (*left*), receives his nomination certificate

In addition to promoting BCI research through the Awards, the Gala Award Ceremonies have also provided a way for BCI experts to relax and interact. The attendees had just seen each other's posters, talks, and demonstrations. Many of them wanted to ask each other for more information, seek research collaborations, or just congratulate them on their good work. This can be difficult during the hectic conference schedule. Indeed, many of the discussions at the 2014 ceremony did lead to new research collaborations. Thus, the annual ceremonies have helped to encourage BCI research, along with the annual Award and book series. The following three pictures present the 2014 Gala Award Ceremony (Figs. 1, 2, 3).

The BCI Book Series

This annual book series is another important component of the Annual Award. Each year, G.TEC produces a book that reviews ten outstanding projects. Each group is responsible for writing a book chapter that presents their work. In addition to the work that was nominated, authors may also present related material, such as new work since their submission. Since the nominees are all from active groups, they often have new work that is worth reading. Each book in the annual book series also includes both introduction and summary chapters.

The book series is not just meant for experts in BCIs, engineering, neuroscience, medicine, or other fields. We editors have asked the authors to present material in a friendly manner, without assuming that the reader has a technical background. We also reviewed each chapter to improve readability. Chapters include figures and references to help support the text and guide readers to additional information. While chapters do introduce some advanced material, this book should be clear to most readers with an interest in BCI research.

This year's nominees, and their corresponding chapters, present a myriad of innovations in different facets of BCIs. The chapters include new invasive and noninvasive improvements, innovative signal processing, novel ways to control output devices, and new ways to interact with the user. For example, one chapter by Rutkowski and colleagues introduces an airborne ultrasonic tactile display, a new approach to touch-based interaction. Two chapters (by Ibáñez, Mrachacz-Kersting, and their colleagues) introduce new ways to use BCI technology to study brain plasticity and improve outcomes when patients conduct motor rehabilitation. Two other chapters (by McMullen, Boulay, and their colleagues) explore new BCIs to improve grasp function and other daily activities. These and other chapters present cutting-edge work that should lead to new directions in BCI research.

Projects Nominated for the BCI Award 2014

This year, the selection process was quite challenging. 69 top-level projects were submitted from around the world. The jury selected the following ten nominees in 2014, presented in alphabetical order:

- P. Brunner[a], K. Dijkstra[a], W. Coon[a], J. Mellinger[a], A. L. Ritaccio[a], G. Schalk[a] ([a]Wadsworth Center and Albany Medical College, US)
  Towards an auditory attention BCI
- J. Gomez-Pilar[a], R. Corralejo[a], D. Álvarez[a], R. Hornero[a] ([a]Biomedical Engineering Group, E. T. S. I. Telecomunicación, University of Valladolid, ES)
  Neurofeedback training by motor imagery based-BCI improves neurocognitive areas in elderly people
- K. Hamada[a], H. Mori[b], H. Shinoda[a], T.M. Rutkowski[b,c] ([a]The University of Tokyo, JP, [b]Life Science Center of TARA, University of Tsukuba, JP, [c]RIKEN Brain Science Institute, JP)
  Airborne ultrasonic tactile display BCI
- J. Ibáñez[a], J. I. Serrano[a], M. D. del Castillo[a], E. Monge[b], F. Molina[b], F.M. Rivas[b], J.L. Pons[a] ([a]Bioengineering Group of the Spanish National Research Council (CSIC), [b]LAMBECOM group, Health Sciences Faculty, Universidad Rey Juan Carlos, Alcorcón, ES)
  Heterogeneous BCI-triggered functional electrical stimulation intervention for the upper-limb rehabilitation of stroke patients
- D. McMullen[a], G. Hotson[b], M. Fifer[c], K. Kaytal[d], B. Wester[d], M. Johannes[d], T. McGee[d], A. Harris[d], A. Ravitz[d], W. S. Anderson[a], N. Thakor[c], N. Crone[a] ([a]Johns Hopkins University School of Medicine Department of Neurology and Neurosurgery, [b]Johns Hopkins University Department of Electrical Engineering,[c]Johns Hopkins University Department of Biomedical Engineering,[d]Johns Hopkins University Applied Physics Laboratory, US)
  Demonstration of a semi-autonomous hybrid brain-machine interface using human intracranial EEG, eye tracking, and computer vision to control a robotic upper limb prosthetic
- K. J. Miller[a], G. Schalk[b], D. Hermes[c], J. G. Ojemann[d], R. P.N. Rao[e] ([a]Department of Neurosurgery, Stanford University, [b]Wadsworth Center and Albany Medical College, [c]Department of Psychology, Stanford University, [d]Department of Neurological Surgery, University of Washington, [e]Department of Computer Science and Engineering, University of Washington, US)
  Unsupervised decoding the onset and type of visual stimuli using electrocorticographic (ECoG) signals in humans
- N. Mrachacz-Kersting[a], N. Jiang[b], S. Aliakbaryhosseinabadi[a] R. Xu[b], L. Petrini[a], R. Lontis[a], M. Jochumsen[a], K. Dremstrup[a], D. Farina[b] ([a]Sensory-Motor Interaction, Department of Health Science and Technology, DK, [b]Dept. Neurorehabilitation Engineering Bernstein Center for Computational Neuroscience University Medical Center, DE)
  The changing brain: bidirectional learning between algorithm and user

- *M. M. Shanechi[a,b], A. L. Orsborn[c], H. G. Moorman[c], S. Gowda[b], S. Dangi[b], J. M. Carmena[b,c] ([a]School of Electrical and Computer Engineering, Cornell University, [b]Department of Electrical Engineering and Computer Sciences, University of California, Berkeley, [c]UC Berkeley UCSF Joint Graduate Program in Bioengineering, US)*
  Rapid control and feedback rates in the sensorimotor pathway enhance neuro-prosthetic control
- *F. R. Willett[a,b,c], H. A. Kalodimos[a,b,c], D. M. Taylor[a,b,c] ([a]Cleveland Clinic, Neurosciences, [b]Cleveland VA Medical Center, [c]Case Western Reserve University, Biomedical Engineering, US)*
  Retraining the brain to directly control muscle stimulators in an upper-limb neuroprosthesis
- *A. Wilson[a], R. Arya[a] ([a]Cincinnati Children's Hospital Medical Center, US)*
  Real-time bedside cortical language mapping during spontaneous conversation with children

Summary

Overall, the annual BCI Awards—along with the book series and ceremonies—have been successful in their main goal: to recognize top BCI research worldwide. They have also helped to bring BCI researchers together, encouraging discussion and collaboration among groups that might not otherwise communicate. These goals are addressed in a quote from Dr. Christoph Guger, CEO at G.TEC, presented on the annual Award websites:

"We have a vital interest in promoting excellence in the field of BCI. Achieving our goal to make BCIs more powerful, more intelligent and more applicable for patients' and care-givers' everyday lives strongly relies on a creative research community worldwide. The Annual BCI Research Award allows us to look back at highlights of BCI research in 20 years and to see how the field changed".

Real-Time Mapping of Natural Speech in Children with Drug-Resistant Epilepsy

Ravindra Arya and J. Adam Wilson

Introduction

Epilepsy is one of the commonest neurological disorders in childhood. Among people with epilepsy, 25–30 % do not respond to treatment with anti-seizure medications (Kwan and Brodie 2000; Mohanraj and Brodie 2006). The probability of achieving lasting seizure remission is ≤10 % in people who have failed two or more appropriately chosen anti-seizure medications used in adequate regimens (Mohanraj and Brodie 2006). Epilepsy surgery is often the only option for long-term seizure control in these people with drug-resistant epilepsy (DRE). Epilepsy surgery requires extensive pre-operative evaluation to determine the anatomical and functional relationships between the seizure-onset zone and cortical areas sub-serving essential sensorimotor and language functions. In a majority of patients, particularly those without an obvious epileptogenic lesion, this evaluation necessitates surgical implantation of subdural electrodes for extended monitoring and functional cortical mapping.

The current method of language mapping in clinical practice involves electrical stimulation of pairs of implanted subdural electrodes using increasing current strengths, as the patient repetitively performs a task related to language production and/or reception. The functional endpoint of the cortical stimulation is to achieve inhibition of the so-called language task, thereby providing localization of cortical areas relevant for language function and their proximity to seizure-onset zone. This procedure of electrical cortical stimulation (ECS) is fraught with multiple safety issues and concerns about ecological validity of the chosen tasks.

R. Arya (✉) · J.A. Wilson
Comprehensive Epilepsy Center, Division of Neurology, MLC 2015, Cincinnati Children's
Hospital Medical Center, 3333 Burnet Avenue, Cincinnati, OH 45229, USA
e-mail: Ravindra.Arya@cchmc.org

© The Author(s) 2015
C. Guger et al. (eds.), *Brain-Computer Interface Research*,
SpringerBriefs in Electrical and Computer Engineering,
DOI 10.1007/978-3-319-25190-5_2

Limitations of Electrical Cortical Stimulation

ECS is a time and resource intensive procedure that entails the risks of stimulating intracranial pain-sensitive structures, triggering electrical after-discharges, and eliciting seizure(s). Occurrence of a triggered seizure precludes further mapping in a given patient, at least temporarily. There have been long standing concerns about the choice of language tasks as well. The commonly used paradigms involve: naming of pictures, objects, or printed words; verb generation in response to an auditory or a visual stimulus; or passive listening to various types of stimuli. However, there is a lack of uniformity in the task selection, and different centers use one or more language tasks in customized protocols with little, if any, external validation (Hamberger 2007). Among the conventional language tasks, visual naming has been often used based on the heuristic that dysnomia is a feature of nearly all aphasia syndromes (Hamberger 2007). However, there have been long-standing concerns regarding its ecological validity, and at a more fundamental level, if such a task can capture the complexity of human language network (Hamberger and Cole 2011).

Finally, the performance of language tasks is a function of patient's intelligence, verbal ability, and capability for sustained participation in a repetitive task, in a rather hostile environment—the epilepsy monitoring unit. In clinical practice, this degree of patient co-operation can be difficult to procure, particularly in children or patients with intellectual disability.

Electrocorticographic Mapping Using Spectral Power Modulation

In view of the challenges and barriers associated with ECS, alternative methods of language mapping were explored. Experiments using event related potentials, and application of signal processing methods to digital EEG, led to recognition of stimulus-locked decreases in power in the α band (event-related desynchronization) and, more importantly, event-related synchronization in the high-γ (≥70 Hz) frequency band (Pfurtscheller 1999). An early study looking at the clinical validity of this approach compared electrodes showing a power increase in 80–100 Hz bands to those testing 'positive' for aphasia (true naming deficit) or oral motor deficit (dysarthria or apraxia severe enough to preclude successful naming) during visual naming of line drawings (Sinai et al. 2005). In 13 adult patients, a sensitivity of 43 % and specificity of 84 % was observed (Sinai et al. 2005). The authors recognized that disruption of oral motor function with ECS at certain electrode locations barred testing for true aphasia at those sites, in addition to causing pain, particularly in posterior temporal locations. While this approach circumvents the safety issues involved with ECS, it still depends on stimulus-response paradigms, which require sustained patient attention and cooperation. As discussed above, there are several potential barriers to this approach in clinical setting, particularly in children.

Electrocorticographic Mapping with Natural Speech

In 2008, Towle et al. described the topography of power changes between 'talking' and 'not talking' epochs, by identifying the times when patient(s) were conversing with the family or staff, on video review (Towle et al. 2008). The maximum increase in high-γ power was noted over the peri-sylvian frontal lobe and posterior superior temporal gyrus in 2/5 patients, which corresponded with activation observed with formal language tasks and also with conventional anatomy of language network. Further, in another study, changes in high-γ power related to retrospectively selected epochs of natural language were compared with conventional ECS in 3 German speaking adults. When including both oral motor and speech sites as true positives, the study reported a high specificity (96.7 %) but very low sensitivity (18.9 %) (Ruescher et al. 2013). The authors concluded that their procedure was not usable for clinical language mapping.

Recently, our group reported using BCI2000 with the Signal Modeling for Real-time Identification and Event Detection (SIGFRIED) procedure (Schalk et al. 2004; Brunner et al. 2009). We first obtained a baseline model of high-γ brain activity for each channel during a 6-min period in a quiet room free from distractions, to the extent possible in a busy hospital environment. Then, during the experiment, the high-γ features from each channel were compared to the baseline model every 100 ms, and a single score was returned estimating the probability that the current task-space sample belongs to the baseline. Thus, periods of activity significantly different from the baseline were obtained for each channel, which were then accumulated for known task periods. Typically, we sequenced these periods, and the subject responded to instruction or stimuli on the computer monitor or speakers. That is, known timestamps for presented stimuli were saved synchronously with the ECoG dataset, allowing statistics to be computed offline at a later time.

Further, we plotted electrodes with significant high-γ modulation on a 3D cortical model derived from the patient's own T1-weighted magnetic resonance imaging (MRI) of the brain. Sensitivity, specificity, positive and negative predictive values (PPV, NPV), and classification accuracy were calculated compared to ECS, including both naming and oral motor deficits as true positives. In the 7 reported patients (4 males, mean age 10.28 ± 4.07 years), significant high-γ responses were observed in classic language areas in the left hemisphere and in some homologous right hemispheric areas. The sensitivity and specificity of ECoG high-γ mapping were 88.89 and 63.64 %, respectively, and PPV and NPV were 35.29 and 96.25 %, with an overall accuracy of 68.24 % (Arya et al. 2015). We believe that this study supports the feasibility of language mapping with ECoG high-γ modulation during spontaneous/natural conversation, and its accuracy compared to traditional ECS.

Methodology of Real-Time Electrocorticographic Natural Speech Mapping

In the above study by our group, the signal processing and localization of the topography of high-γ modulation was carried out offline after the experiment (Arya et al. 2015). However, we have automated this pipeline to achieve real-time visualization of activation associated with natural speech, allowing research outcomes to be available at the bedside. We have been using these tools in an ongoing study.

Participants and experimental protocol

All patients who are undergoing pre-surgical evaluation with implanted subdural electrodes in the epilepsy monitoring unit at Cincinnati Children's Hospital Medical Center, and are able to converse with the investigator, are eligible for inclusion. A baseline period of ECoG signals is captured for each participant in an awake, relaxed and silent state in a quiet environment. Then, age-appropriate open-ended questions are used to converse with the participant. The participants do not have any prior knowledge of these questions. In addition, we also record ECoG during conventional stimulus-response tasks including overt and covert visual naming, story listening, and music listening. ECS is performed extra-operatively based on clinical need entirely. The selection of electrode pairs, stimulation end-point, and protocol for visual naming are at the discretion of the neurologist carrying out the ECS.

Data Collection

ECoG signals are split at the break-out box, amplified and digitized with a g. USBamp (g.tec Medical Engineering, Austria) amplifier, controlled by a computer running the BCI2000 software system. ECoG data is obtained at a sampling rate of 1.2 kHz (Fig. 1). Audio is captured using a wireless microphone system, with the patient and doctor each wearing a Shure cardioid lavalier microphone. Audio gain levels were adjusted as appropriate, and an audio mixer board was used to separate each audio source to the left (patient) and right (doctor) channels. The audio was recorded to the computer in a .wav file during the experiment using a BCI2000 module. Finally, the left and right audio streams were also split to a g.tec Trigger Box (g.tec Medical Engineering, Austria), which converts an input signal to a digital transistor-transistor logic (TTL) pulse when the signal crosses an adjustable threshold. These two digital signals were recorded by the g.tec amplifier synchronously with the ECoG data, allowing the onset of speech for the doctor and patient each to be measured with sub-millisecond precision, and any timing inaccuracies in the .wav file to be corrected offline for detailed analysis.

3D Cortical Model

For each patient, a T1-weighted spin-echo MRI of brain is co-registered with a head computed tomographic scan (CT) with Curry scan 7 software (Compumedics, Charlotte, NC), and segmented 3D cortical models are exported from Curry. A custom BCI2000 module based on the original SIGFRIED was written to display the 3D model and activation projections. The stereotactic coordinates of each electrode are obtained and transformed to the Talairach space.

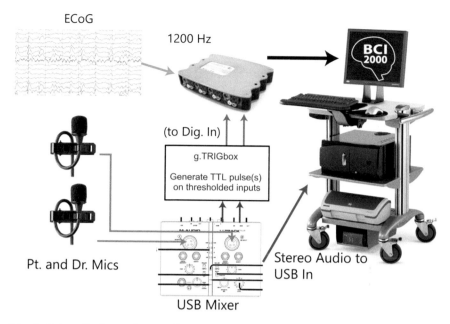

Fig. 1 Data collection and processing for real-time language mapping (see text for further explanation)

Data Analysis

The signal processing approach is essentially similar to that used in our spontaneous conversation mapping study (Arya et al. 2015). Briefly, a statistical model of the quiet baseline data is generated using the SIGFRIED procedure. During the conversation, SIGFRIED is again used to identify electrodes that show a statistically significant change from the baseline model, for the 70–116 Hz frequency band. The SIGFRIED method easily lends itself to real-time functional mapping, as demonstrated both for motor and language cortices (Brunner et al. 2009). Also, it requires no a priori information about the expected modulated frequency bands, since any changes are encapsulated in a single score, and no subjective assumptions about the data significance are required.

We compared signal activities in two ways: expressive speech (patient speaking) vs. baseline, and receptive language (patient listening/investigator speaking) versus baseline. The audio digital triggers provided the task conditions for the listening and speaking during the experiment, thereby allowing our visualization routines to accumulate the statistics for listening versus baseline, and speaking versus baseline periods. Importantly, the stimulus onsets and durations are known beforehand in typical experiments, but the stimulus onsets and durations were calculated 'on the fly' using the digital audio triggers in our experiments. We then compare SIGFRIED scores during these 2 conditions, using Student's t-tests with Bonferroni

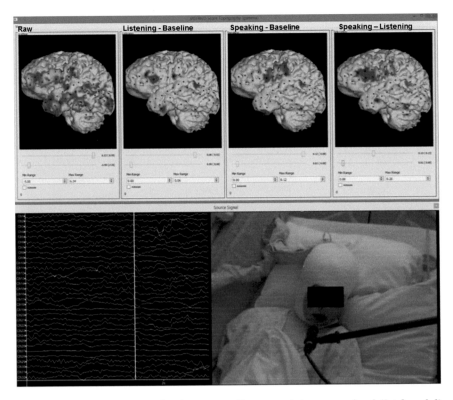

Fig. 2 Real-time language mapping in progress. *Upper panel* shows raw signal (1st from *left*), and averaged signal for listening (2nd from *left*) and speaking (3rd from *left*) compared to baseline. Note the activation in posterior superior temporal gyrus during listening and in Broca's and oral motor areas during speech. Also notice activation in the tip of the temporal lobe

correction based on the number of channels. For plotting activations on the 3D model of cortex, the negative common logarithm of the p-values $[-\log_{10}(p)]$ is used. This metric varies directly with cortical activation, is easy to interpret, and additive, such that it can be used to show the average brain activation across several subjects. The process is automated and the activation plot for the individual patient is available for visualization in real time (Fig. 2).

Comparisons

During extra-operative ECS, the electrodes found to serve language or oral motor functions are regarded as true positives (ECS+), whereas other electrodes where stimulation was carried out without any such response are regarded as true negatives (ECS−). The electrodes that showed significant high-γ modulation relative to baseline (HGM+) are selected by taking the largest $-\log_{10}p$ value between the language and baseline conditions $[-\log_{10}(p) \geq -\log_{10}(0.01/N)$, where N is the number of ECS electrodes]. Sensitivity, specificity, positive and negative predictive values, and classification accuracy will be calculated to validate HGM against ECS.

Results

Only the electrodes tested during clinical ECS are analyzed. For all comparisons, ECS electrodes with either naming (aphasia/paraphasic errors) or oral motor deficits were deemed as true positives. In the 3 patients (2 left and 1 right hemispheric subdural electrodes) recorded so far, real-time mapping of overt naming was found to have a high NPV (93.6 %, 95 % CI 82.5–98.7 %), and moderate sensitivity (75.0 %, 95 % CI 42.8–94.5 %) and specificity (75.9 %, 95 % CI 62.8–86.1 %) compared to ECS (Table 1). In contrast, covert naming was found to have a high specificity (98.3 %, 95 % CI 90.7–100.0 %) and NPV (85.1 %, 95 % CI 74.3–92.6 %), but poor sensitivity (16.7 %, 95 % CI 2.1–48.4 %, Table 1).

Natural speech mapping was also found to have a high NPV (95.5 %, 95 % CI 84.5–99.4 %) and sensitivity (84.6 %, 95 % CI 54.6–98.1 %) with moderate specificity (65.6 %, 95 % CI 52.7–77.1 %, Table 1).

Significance and Conclusions

Using the above methodology, we found a high NPV with either task-based or natural speech mapping. This implies that the electrode locations that this method classified as not participating in language networks are truly unlikely to have language function, which is the critical information for neurosurgical decision making. One of the limitations of this study is that the physiological role of those electrodes that were not tested by ECS cannot be determined.

We believe that this methodology has the potential to obviate the risks associated with ECS including adverse effects of premedication, risk of iatrogenic stimulated seizures, and pain from trigeminal afferent stimulation. Application of this technology to patient care is likely to save valuable time by circumventing ECS, or at a minimum, informing the conventional mapping by preselecting the electrodes showing activation. This can be crucial in children who can only cooperate with formal testing for a limited time. In future, we expect this technique to provide the basis for modeling coherence and causal relationships between components of the cortical networks engaged in cognitive processing. This will likely enable better understanding of dynamic spatial and temporal relationships between cortical areas involved in language function, compared to the dichotomous results of electrical stimulation. ECS merely determines whether there was response inhibition at a particular electrode contact or not, and does not provide any knowledge about network dynamics.

Table 1 Comparison of real-time task based and spontaneous natural speech mapping with conventional electrical cortical stimulation

Overt naming

Patient Identifier	Hemisphere	ECS+ |PN+	ECS+ |PN−	ECS− |PN+	ECS− |PN−	Validation statistics (95 % CI)
BD	Left	1	1	5	12	Sensitivity 75.0 % (42.8–94.5 %)
BG	Right	2	1	7	19	Specificity 75.9 % (62.8–86.1 %)
CA	Left	6	1	2	13	PPV 39.1 % (19.7–61.5 %)
Total		9	3	14	44	NPV 93.6 % (82.5–98.7 %)

Covert naming

Patient identifier	Hemisphere	ECS+ |Cov+	ECS+ |Cov−	ECS− |Cov+	ECS− |Cov−	Validation statistics (95 % CI)
BD	Left	0	2	1	16	Sensitivity 16.7 % (2.1–43.4 %)
BG	Right	2	1	0	26	Specificity 98.3 % (90.7–100.0 %)
CA	Left	0	7	0	15	PPV 66.7 % (9.4–99.2 %)
Total		2	10	1	57	NPV 85.1 % (74.3–92.6 %)

Spontaneous speech

Patient identifier	Hemisphere	ECS+ |Sp+	ECS+ |Sp−	ECS− |Sp+	ECS− |Sp−	Validation statistics (95 % CI)
BD	Left	3	0	3	15	Sensitivity 84.6 % (54.6–98.1 %)
BG	Right	3	0	17	14	Specificity 65.6 % (52.7–77.1 %)
CA	Left	5	2	2	13	PPV 33.3 % (18.0–51.8 %)
Total		11	2	22	42	NPV 95.5 % (84.5–99.4 %)

References

R. Arya, J.A. Wilson, J. Vannest et al., Electrocorticographic language mapping in children by high-gamma synchronization during spontaneous conversation: comparison with conventional electrical cortical stimulation. Epilepsy Res. **110**, 78–87 (2015)

P. Brunner, A.L. Ritaccio, T.M. Lynch et al., A practical procedure for real-time functional mapping of eloquent cortex using electrocorticographic signals in humans. Epilepsy Behav. **15**, 278–286 (2009)

M.J. Hamberger, Cortical language mapping in epilepsy: a critical review. Neuropsychol. Rev. **17**, 477–489 (2007)

M.J. Hamberger, J. Cole, Language organization and reorganization in epilepsy. Neuropsychol. Rev. **21**, 240–251 (2011)

P. Kwan, M.J. Brodie, Early identification of refractory epilepsy. New Engl. J. Med. **342**, 314–319 (2000)

R. Mohanraj, M.J. Brodie, Diagnosing refractory epilepsy: response to sequential treatment schedules. Eur. J. Neurol. **13**, 277–282 (2006)

G. Pfurtscheller, Lopes da Silva FH. Event-related EEG/MEG synchronization and desynchronization: basic principles. Clin. Neurophysiol. **110**, 1842–1857 (1999)

J. Ruescher, O. Iljina, D.M. Altenmuller, A. Aertsen, A. Schulze-Bonhage, T. Ball, Somatotopic mapping of natural upper- and lower-extremity movements and speech production with high gamma electrocorticography. NeuroImage **81**, 164–177 (2013)

G. Schalk, D.J. McFarland, T. Hinterberger, N. Birbaumer, J.R. Wolpaw, BCI2000: a general-purpose, brain-computer interface (BCI) system. IEEE Trans. Biomed. Eng. **51**, 1034–1043 (2004)

A. Sinai, C.W. Bowers, C.M. Crainiceanu et al., Electrocorticographic high gamma activity versus electrical cortical stimulation mapping of naming. Brain **128**, 1556–1570 (2005)

V.L. Towle, H.A. Yoon, M. Castelle et al., ECoG gamma activity during a language task: differentiating expressive and receptive speech areas. Brain **131**, 2013–2027 (2008)

Brain-Computer Interfaces for Communication and Rehabilitation Using Intracortical Neuronal Activity from the Prefrontal Cortex and Basal Ganglia in Humans

Chadwick B. Boulay and Adam J. Sachs

Brain-computer interfaces (BCIs) can help individuals with central nervous system (CNS) deficits recover lost function by enabling new communication channels (Wolpaw et al. 2002; Hochberg et al. 2012; Collinger et al. 2013) or by inducing and guiding adaptive plasticity for rehabilitation (Daly and Sitaram 2012; Mukaino et al. 2014).

BCIs for communication—Ongoing research in BCIs for communication aims to maximize the rate of information transfer from the brain to the computer. The majority of BCIs for communication are driven by brain signals that were previously known to be responsive to changes in user intention (Farwell and Donchin 1988; Pfurtscheller et al. 1997; Birbaumer et al. 2000; Allison et al. 2008). The neurophysiology underlying the brain signal and its relation to the condition necessitating the BCI are secondary to the need for a brain signal that is accessible, robust, and intuitive to control. There are only a few brain signals that meet these criteria.

BCIs for rehabilitation—Ongoing research in BCIs for rehabilitation aims to maximize recovery of function after CNS trauma or disease. There are at least two mechanisms that might enhance rehabilitation through continued BCI use. First, a BCI for rehabilitation might induce Hebbian-like plasticity through co-activation of motor areas in the brain and the periphery by using the detection of motor-related brain signals to trigger peripheral sensorimotor stimulation (Mrachacz-Kersting et al. 2012; Mukaino et al. 2014). Second, a BCI for rehabilitation might facilitate recovery through operant conditioning of brain signals to push the CNS toward a

C.B. Boulay (✉) · A.J. Sachs
Department of Surgery, Division of Neurosurgery, The Ottawa Hospital Research Institute,
Ottawa, Canada
e-mail: cboulay@uottawa.ca

A.J. Sachs
e-mail: asachs@toh.on.ca

C.B. Boulay · A.J. Sachs
Faculty of Medicine, The University of Ottawa, Ottawa, Canada

© The Author(s) 2015
C. Guger et al. (eds.), *Brain-Computer Interface Research*,
SpringerBriefs in Electrical and Computer Engineering,
DOI 10.1007/978-3-319-25190-5_3

19

state that is more permissive to traditional interventions (Pichiorri et al. 2015). In either case, the neurophysiology underlying the CNS deficit should be the primary factor in the selection of the BCI signal source. However, most BCIs for rehabilitation use the same brain signals as BCIs for communication, without much attention to the subject-specific deficits. More work is needed to identify optimal brain signals for rehabilitative BCIs.

Alternative BCI signal sources—The development of real-time functional magnetic resonance imaging (rtfMRI) has made it possible to use the metabolic signals from almost any brain region to drive a BCI (Sulzer et al. 2013). rtfMRI is less practical for communication, but it shows great potential for inducing plasticity in BCIs for rehabilitation (Shibata et al. 2011; Ruiz et al. 2013). Electrocorticography (ECoG) in patients with intractable epilepsy provides access to brain signals that are unattainable with less-invasive methods. ECoG electrode placement is dictated by clinical need and as a result most ECoG BCI studies used temporal and motor cortices (Leuthardt et al. 2004; Schalk and Leuthardt 2011); we identified one study that used prefrontal cortex (PFC) (Vansteensel et al. 2010). Recently, ECoG electrodes have been implanted in motor cortex exclusively for BCI purposes (Wang et al. 2013). Finally, a few researchers are chronically implanting microelectrodes into the brains of severely disabled participants to drive BCIs with neuronal spiking in motor (Hochberg et al. 2006; Collinger et al. 2013) and parietal cortices (Aflalo et al. 2015).

There are yet other brain signal sources for BCIs that are apt for investigation. For example, the clinical procedure to implant deep brain stimulation (DBS) electrodes for the treatment of mood and motor disorders exposes PFC and records from microelectrodes in the basal ganglia (BG). PFC and BG may be useful signal sources for BCIs. At the Ottawa Hospital, we surgically implant DBS electrodes into the BG and thalamus for the treatment of Parkinson's disease (PD), dystonia, and essential tremor. We take these opportunities to investigate neuronal activity in the prefrontal cortex and basal ganglia as potential signal sources in BCIs for communication and rehabilitation.

Our Research Platform

Participants—Our participants are from a group of individuals diagnosed with Parkinson's disease (PD), dystonia, essential tremor, or other movement disorders who have been selected to undergo surgical implantation of deep brain stimulation (DBS) electrodes as part of their standard-of-care clinical treatment. The institutional research ethics board has approved this research project, and all participants provide informed consent.

Brain signal acquisition—During the DBS electrode implantation procedure, neuronal spiking activity and LFPs are monitored from multiple microelectrodes as they are driven along a linear trajectory from the cortical surface down to the

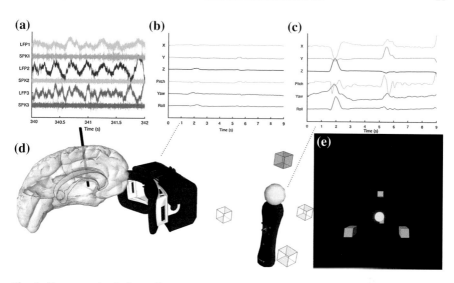

Fig. 1 Our research platform allows us to record simultaneously (panel **a**) intracortical and/or subcortical neurophysiological activity, (panel **b**) the head pose, and (panel **c**) the pose of an effector held firmly in the hand. The participant sees in stereoscopic 3D four virtual targets and a sphere representing the effector (a monocular screenshot is shown in panel **e**), and the colour of the sphere is controlled by brain signal features extracted from the neurophysiological data in real-time

stimulation target. Microelectrode insertion is guided by a Microdrive secured to the skull using a clinical frameless stereotaxy platform (Nexframe); the participants are awake and may move their head freely. We pause driving the microelectrodes along the target trajectory to perform the BCI experiment. We record the signals from the output of the microelectrode recording system into a separate computer running the experimental software that processes the brain signals and presents the task through an application using a 3D game engine.

Behavioural measurements—The application coordinates inputs from an optical hand tracker and a head tracker and displays a representation of the participant's hand as a sphere in a virtual workspace. The application provides visual feedback through a consumer-grade virtual reality head-mounted display that presents the sphere so it is perceived to be in the participant's true hand location. The head pose (position and orientation), the hand pose, and the task state are continuously stored to a data file (Fig. 1).

BCI tasks—The BCI experiment follows the general design of our previously successful paradigm (Boulay et al. 2011) and comprises three tasks. In the first task, we identify brain signals that predict behavioural performance. Participants are cued to perform a visually guided centre-out task to different locations in space indicated by virtual targets. We then analyze the recorded data to identify linear combinations of brain signal features that correlate with motor performance metrics including direction, reaction time, smoothness, and error. Extracted brain signal features include spectral band-power (especially in the beta band), neuronal spiking rates, and phase-amplitude coupling.

In the second task, we train participants to modulate the brain signal feature identified during the first task. For example, the magnitude of the feature is transformed online into the colour of the sphere representing hand location, and participants are cued to change the colour of the sphere to match the indicated colour.

In the third task, we examine the interaction between volitional modulation of brain signals and behavioural performance. For example, participants must change the colour of the sphere by modulating their brain signals, then they must move their hand to the target location. We examine if there are differences in movement performance between trials in which the brain signal predicted good performance and trials in which the brain signal predicted poor performance.

Prefrontal Cortex as a BCI Signal Source

The PFC is known to be involved in decision-making and goal-directed behavior, and might also be a reliable signal source for goal-selection BCIs. Previous single unit studies have implicated the PFC in attention (Everling et al. 2002; Lennert and Martinez-Trujillo 2013), working memory (Jacobson 1936; Fuster and Alexander 1971; Kubota and Niki 1971; Goldman-Rakic 1995; Miller et al. 1996; Mendoza-Halliday et al. 2014), task rule representation (White and Wise 1999; Wallis et al. 2001; Bongard and Nieder 2010; Buschman et al. 2012) and the integration of information for decision making (Kim and Shadlen 1999). It might therefore be possible to decode behavioural goals, i.e. decision outcomes, from PFC neural activity.

We are beginning to examine the PFC as a BCI signal source in primates (humans and monkeys). We have an ongoing collaboration with Dr. Julio Martinez-Trujillo at the University of Western Ontario to investigate this possibility in monkeys. Thus far, we have demonstrated that we could decode the direction of covert attention from single realizations of lateral PFC neuronal activity (Tremblay et al. 2015), and we are working toward decoding decision outcomes and the context underlying the decisions. In humans, we are progressing toward implanting microelectrodes in the PFC of severely disabled individuals.

The Subthalamic Nucleus as a BCI Signal Source

The basal ganglia are part of a distributed cortico-basal ganglia-thalamo-cortical loop involved in voluntary movements (Wichmann et al. 1994), eye movements (Hikosaka et al. 2000), and cognition (Middleton and Strick 2000). A primary control structure in this network is the sub-thalamic nucleus (STN) (Parent and Hazrati 1995). STN activity is modulated during passive movement of the limbs as well as with onset of volitional movements, and the neurons have a dense somatotopy with little functional overlap (Wichmann et al. 1994). A small number of electrodes in the STN could sample from functionally discrete neurons that are modulated differently for different tasks, and could thus be used to drive a multi-dimensional BCI for communication. However, it is unknown if STN activity can be modulated volitionally in the absence of overt movements, such as during motor imagery or motor intention in disabled individuals.

Several movement disorders are caused by disruption to the basal ganglia. For example, Parkinson's disease (PD) is characterized by motor deficits due to disrupted transmission in the basal ganglia (BG)–thalamus–cortex circuit (Brown

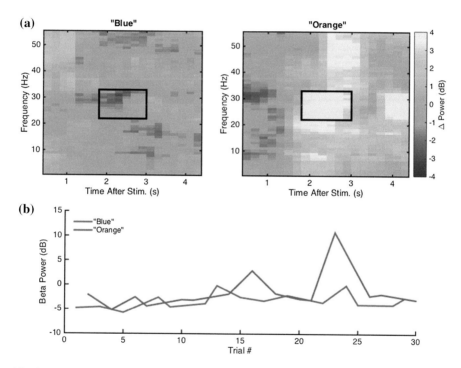

Fig. 2 We have evidence from one participant that patients with Parkinson's disease can acquire some volitional control over brain signal features that are implicated in Parkinsonian symptom severity. A participant was instructed to imagine fluid movements when receiving the *Blue* cue and to imagine akinesia when receiving the *Orange* cue while keeping their hands at rest on their lap. Sub-thalamic nucleus beta-band power was transformed in real-time to the colour of an immobile sphere in the center of the participant's field of view. By the final 14 trials of training, there were differences in the STN spectrogram between *Orange* and *Blue* trials (**a**). The beta-band power (time and frequency ranges indicated with *black rectangles* in panel **a**) initially did not differ between cue conditions but, over the course of 30 feedback trials, the participant was able to achieve some volitional beta synchronization after receiving the *Orange* cue (**b**)

2003). This circuit exhibits pathalogical neuronal synchronization in the beta frequency band (13–30 Hz) proportional to motor deficit severity, and beta synchronization is decreased during treatment by levodopa or DBS in the STN (Kühn et al. 2006; Jenkinson and Brown 2011). The magnitude of phase amplitude coupling (PAC) between beta oscillatory phase and neuronal spiking in the STN is also predictive of PD symptoms (Yang et al. 2014). Levodopa and STN DBS improve the fidelity of the BG-thalamus-cortex circuit by providing extrinsic inputs that disrupt the pathological synchronization and PAC. However, these treatments have side effects, have limited efficacy, and are burdensome. Alternatively, it may be possible to disrupt the pathological synchronization of the BG-thalamus-cortex circuit through long-term intrinsic circuit modification. A BCI for rehabilitation could be used to condition beta power (and/or PAC) to more normal levels with concomitant reduction in PD symptom severity.

Thus far we have identified correlations between STN beta oscillations and movement performance (Fig. 2). Using a single electrode, we decoded hand posture with sufficient accuracy for binary communication and we identified signal features that predict movement direction. Our preliminary results also demonstrate that humans may acquire some volitional control of basal ganglia beta oscillations.

Outlook

We expect our research program to produce findings that contribute to the development of both BCIs for communication and for rehabilitation. Our research using the PFC as a BCI signal source should help identify neural correlates of decision outcomes that can then be used to drive a goal-selection BCI for communication. Our research in the basal ganglia will help determine if these signal sources are reliable and intuitive in BCI operation. Further, our basal ganglia research will motivate the development of devices for long-term BCI training to alleviate motor deficits in disorders such as dystonia and PD.

Our research platform also enables other neuroscientific inquiry owing to the rich data it provides. The combined head pose, hand pose, visual stimulus information, and intracortical neurophysiological data allow unique investigations into the roles of the PFC and the different basal ganglia structures in goal-oriented behaviour. For example, with these data we can investigate the neuronal representation of physical space (Sajad et al. 2014). By manipulating the virtual reality feedback we can dissociate the influence of visual and proprioceptive feedback on neuronal activity and we can investigate changes in neuronal activity as participants learn new arbitrary rules governing the virtual world.

References

T. Aflalo, S. Kellis, C. Klaes, B. Lee, Y. Shi, K. Pejsa, K. Shanfield, S. Hayes-Jackson, M. Aisen, C. Heck, C. Liu, R.A. Andersen, Decoding motor imagery from the posterior parietal cortex of a tetraplegic human. Science **80**(348), 906–910 (2015)

B.Z. Allison, D.J. McFarland, G. Schalk, S.D. Zheng, M.M. Jackson, J.R. Wolpaw, Towards an independent brain-computer interface using steady state visual evoked potentials. Clin. Neurophysiol. **119**, 399–408 (2008)

N. Birbaumer, A. Kübler, N. Ghanayim, T. Hinterberger, J. Perelmouter, J. Kaiser, I. Iversen, B. Kotchoubey, N. Neumann, H. Flor, The thought translation device (TTD) for completely paralyzed patients. IEEE Trans. Rehabil. Eng. **8**, 190–193 (2000)

S. Bongard, A. Nieder. Basic mathematical rules are encoded by primate prefrontal cortex neurons. Proc. Natl. Acad. Sci. **107**, 2277–82 (2010)

C.B. Boulay, W.A. Sarnacki, J.R. Wolpaw, D.J. McFarland, Trained modulation of sensorimotor rhythms can affect reaction time. Clin. Neurophysiol. **122**, 1820–1826 (2011)

P. Brown, Oscillatory nature of human basal ganglia activity: relationship to the pathophysiology of Parkinson's disease. Mov. Disord. **18**, 357–363 (2003)

T.J. Buschman, E.L. Denovellis, C. Diogo, D. Bullock, E.K. Miller. Synchronous oscillatory neural ensembles for rules in the prefrontal cortex. Neuron **76**, 838–46 (2012)

J.L. Collinger, B. Wodlinger, J.E. Downey, W. Wang, E.C. Tyler-Kabara, D.J. Weber, A. J. McMorland, M. Velliste, M.L. Boninger, A.B. Schwartz, High-performance neuroprosthetic control by an individual with tetraplegia. Lancet **381**, 557–564 (2013)

J.J. Daly, R. Sitaram, BCI therapeutic applications for improving brain function, in *Brain-Computer Interfaces: Principles and Practice*, ed. by J.R. Wolpaw, E.W. Wolpaw (Oxford University Press, Oxford, 2012), pp. 351–362

S. Everling, C.J. Tinsley, D. Gaffan, J. Duncan, Filtering of neural signals by focused attention in the monkey prefrontal cortex. Nat. Neurosci. **5**, 671–676 (2002)

L.A. Farwell, E. Donchin, Talking off the top of your head: toward a mental prosthesis utilizing event-related brain potentials. Electroencephalogr. Clin. Neurophysiol. **70**, 510–523 (1988)

J.M. Fuster, G.E. Alexander, Neuron activity related to short-term memory. Science **173**, 652–654 (1971)

P.S. Goldman-Rakic, Cellular basis of working memory. Neuron **14**, 477–485 (1995)

O. Hikosaka, Y. Takikawa, R. Kawagoe, Role of the basal ganglia in the control of purposive saccadic eye movements. Physiol. Rev. **80**, 953–978 (2000)

L.R. Hochberg, D. Bacher, B. Jarosiewicz, N.Y. Masse, J.D. Simeral, J. Vogel, S. Haddadin, J. Liu, S.S. Cash, P. van der Smagt, J.P. Donoghue, Reach and grasp by people with tetraplegia using a neurally controlled robotic arm. Nature **485**, 372–375 (2012)

L.R. Hochberg, M.D. Serruya, G.M. Friehs, J.A. Mukand, M. Saleh, A.H. Caplan, A. Branner, D. Chen, R.D. Penn, J.P. Donoghue, Neuronal ensemble control of prosthetic devices by a human with tetraplegia. Nature **442**, 164–171 (2006)

C.F. Jacobson, Studies of cerebral functions in primates: I. The functions of the frontal association areas in monkeys. Comp. Psychol. Monogr. **13**, 1–30 (1936)

N. Jenkinson, P. Brown, New insights into the relationship between dopamine, beta oscillations and motor function. Trends Neurosci. **34**, 611–618 (2011)

J.N. Kim, M.N. Shadlen, Neural correlates of a decision in the dorsolateral prefrontal cortex of the macaque. Nat. Neurosci. **2**, 176–185 (1999)

K. Kubota, H. Niki, Prefrontal cortical unit activity and delayed alternation performance in monkeys. J. Neurophysiol. **34**, 337–347 (1971)

A. A. Kühn, A. Kupsch, G.-H. Schneider, P. Brown, Reduction in subthalamic 8-35 Hz oscillatory activity correlates with clinical improvement in Parkinson's disease. Eur. J. Neurosci. **23**, 1956–1960 (2006)

T. Lennert, J.C. Martinez-Trujillo, Prefrontal neurons of opposite spatial preference display distinct target selection dynamics. J. Neurosci. **33**, 9520–9529 (2013)

E.C. Leuthardt, G. Schalk, J.R. Wolpaw, J.G. Ojemann, D.W. Moran, A brain-computer interface using electrocorticographic signals in humans. J. Neural Eng. **1**, 63–71 (2004)

D. Mendoza-Halliday, S. Torres, J.C. Martinez-Trujillo, Sharp emergence of feature-selective sustained activity along the dorsal visual pathway. Nat. Neurosci. **17**, 1255–1262 (2014)

F. Middleton, P. Strick, Basal ganglia output and cognition: evidence from anatomical, behavioral, and clinical studies. Brain Cogn. (2000)

E.K. Miller, C.A. Erickson, R. Desimone, Neural mechanisms of visual working memory in prefrontal cortex of the macaque. J. Neurosci. **16**, 5154–5167 (1996)

N. Mrachacz-Kersting, S.R. Kristensen, I.K. Niazi, D. Farina, Precise temporal association between cortical potentials evoked by motor imagination and afference induces cortical plasticity. J. Physiol. **590**, 1669–1682 (2012)

M. Mukaino, T. Ono, K. Shindo, T. Fujiwara, T. Ota, A. Kimura, M. Liu, J. Ushiba, Efficacy of brain-computer interface-driven neuromuscular electrical stimulation for chronic paresis after stroke. J. Rehabil. Med. **46**, 378–382 (2014)

A. Parent, L. Hazrati, Functional anatomy of the basal ganglia. ii. the place of subthalamic nucleus and external pallidium in basal ganglia circuitry. Brain Res. Rev (1995)

G. Pfurtscheller, C. Neuper, D. Flotzinger, M. Pregenzer, EEG-based discrimination between imagination of right and left hand movement. Electroencephalogr. Clin. Neurophysiol. **103**, 642–651 (1997)

F. Pichiorri, G. Morone, M. Petti, J. Toppi, I. Pisotta, M. Molinari, S. Paolucci, M. Inghilleri, L. Astolfi, F. Cincotti, D. Mattia, Brain-computer interface boosts motor imagery practice during stroke recovery. Ann. Neurol. **77**, 851–865 (2015)

S. Ruiz, S. Lee, S.R. Soekadar, A. Caria, R. Veit, T. Kircher, N. Birbaumer, R. Sitaram, Acquired self-control of insula cortex modulates emotion recognition and brain network connectivity in schizophrenia. Hum. Brain Mapp. **34**, 200–212 (2013)

A. Sajad, M. Sadeh, G.P. Keith, X. Yan, H. Wang, J.D. Crawford, Visual-motor transformations within frontal eye fields during head-unrestrained gaze shifts in the monkey. Cereb. Cortex (2014). doi:10.1093/cercor/bhu279

G. Schalk, E.C. Leuthardt, Brain computer interfaces using electrocorticographic (ECoG) signals. IEEE Rev. Biomed. Eng. **4**, 140–154 (2011)

K. Shibata, T. Watanabe, Y. Sasaki, M. Kawato, Perceptual learning incepted by decoded fMRI neurofeedback without stimulus presentation. Science **334**, 1413–1415 (2011)

J. Sulzer, S. Haller, F. Scharnowski, N. Weiskopf, N. Birbaumer, M.L. Blefari, A.B. Bruehl, L.G. Cohen, R.C. DeCharms, R. Gassert, R. Goebel, U. Herwig, S. LaConte, D. Linden, A. Luft, E. Seifritz, R. Sitaram, Real-time fMRI neurofeedback: progress and challenges. Neuroimage **76**, 386–399 (2013)

S. Tremblay, F. Pieper, A. Sachs, J.C. Martinez-Trujillo, Attentional filtering of visual information by neuronal ensembles in the primate lateral prefrontal cortex. Neuron **85**, 202–215 (2015)

M.J. Vansteensel, D. Hermes, E.J. Aarnoutse, M.G. Bleichner, G. Schalk, P.C. van Rijen, F.S.S. Leijten, N.F. Ramsey, Brain-computer interfacing based on cognitive control. Ann. Neurol. **67**, 809–816 (2010)

J.D. Wallis, K.C. Anderson, E.K. Miller. Single neurons in prefrontal cortex encode abstract rules. Nature **411**, 953–6 (2001)

W. Wang, J.L. Collinger, A.D. Degenhart, E.C. Tyler-Kabara, A.B. Schwartz, D.W. Moran, D. J. Weber, B. Wodlinger, R.K. Vinjamuri, R.C. Ashmore, J.W. Kelly, M.L. Boninger, An electrocorticographic brain interface in an individual with tetraplegia. PLoS ONE **8**, e55344 (2013)

I.M. White, S. P. Wise. Rule-dependent neuronal activity in the prefrontal cortex. Exp. Brain Res. **126**, 315–335 (1999)

T. Wichmann, H. Bergman, M.R. DeLong, The primate subthalamic nucleus. I. Functional properties in intact animals. J. Neurophysiol. **72**, 494–506 (1994)

J.R. Wolpaw, N. Birbaumer, D.J. McFarland, G. Pfurtscheller, T.M. Vaughan, Brain-computer interfaces for communication and control. Clin. Neurophysiol. **113**, 767–791 (2002)

A.I. Yang, N. Vanegas, C. Lungu, K.A. Zaghloul, Beta-coupled high-frequency activity and beta-locked neuronal spiking in the subthalamic nucleus of Parkinson's disease. J. Neurosci. **34**, 12816–12827 (2014)

Towards an Auditory Attention BCI

Peter Brunner, Karen Dijkstra, William G. Coon, Jürgen Mellinger,
Anthony L. Ritaccio and Gerwin Schalk

Abstract People affected by severe neuro-degenerative diseases (e.g., late-stage amyotrophic lateral sclerosis (ALS) or locked-in syndrome) eventually lose all muscular control and are no longer able to gesture or speak. For this population, an auditory BCI is one of only a few remaining means of communication. All currently used auditory BCIs require a relatively artificial mapping between a stimulus and a communication output. This mapping is cumbersome to learn and use. Recent studies suggest that electrocorticographic (ECoG) signals in the gamma band (i.e., 70–170 Hz) can be used to infer the identity of auditory speech stimuli, effectively removing the need to learn such an artificial mapping. However, BCI systems that use this physiological mechanism for communication purposes have not yet been described. In this study, we explore this possibility by implementing a BCI2000-based real-time system that uses ECoG signals to identify the attended speaker.

Introduction

People affected by severe neuro-degenerative diseases (e.g., late-stage amyotrophic lateral sclerosis (ALS) or locked-in syndrome) eventually lose all muscular control and are no longer able to gesture or speak. They also cannot use traditional assistive

P. Brunner · K. Dijkstra · W.G. Coon · G. Schalk
New York State Department of Health, Center for Adaptive Neurotechnology,
Wadsworth Center, Albany, NY, USA

P. Brunner · A.L. Ritaccio · G. Schalk (✉)
Department of Neurology, Albany Medical College, Albany, NY, USA
e-mail: gerwin.schalk@health.ny.gov

J. Mellinger
Institute of Medical Psychology and Behavioral Neurobiology, University of Tübingen,
Tübingen, Germany

K. Dijkstra
Department of Artificial Intelligence, Donders Institute for Brain, Cognition and Behaviour,
Nijmegen, The Netherlands

© The Author(s) 2015
C. Guger et al. (eds.), *Brain-Computer Interface Research*,
SpringerBriefs in Electrical and Computer Engineering,
DOI 10.1007/978-3-319-25190-5_4

29

communication devices that depend on muscle control, nor typical brain-computer-interfaces (BCIs) that depend on visual stimulation or feedback (Wolpaw et al. 2002; Brunner et al. 2010; Brunner and Schalk 2010). For this population, auditory (Belitski et al. 2011; Furdea et al. 2009; Klobassa et al. 2009; Halder et al. 2010; Schreuder 2010) and tactile BCIs (Brouwer 2010; van der Waal et al. 2012) are two of only a few remaining means of communication (see Riccio et al. 2012 for review).

While visual BCIs typically preserve the identity between the stimulus (e.g., a highlighted 'A') and the symbol the user wants to communicate (e.g., the letter 'A'), all currently used auditory or tactile BCIs require a relatively artificial mapping between a stimulus (e.g., a particular but arbitrary sound) and a communication output (e.g., a particular letter or word). This mapping is easy to learn when there are only few possible outputs (e.g., a yes or no command). However, with an increasing number of possible outputs, such as with a spelling device, this mapping becomes arbitrary and complex. This makes most current auditory and tactile BCI systems cumbersome to learn and use.

Two avenues are being investigated to overcome this limitation. The first avenue is to directly decode expressive silent speech without requiring any external stimuli. In this approach, linguistic elements at different levels (e.g., phonemes, syllables, words and phrases) are first decoded from brain signals and then synthesized into speech. While recent studies have demonstrated this possibility (Pei et al. 2011, 2012; Leuthardt et al. 2011; Martin et al. 2014; Lotte et al. 2015), even invasive brain imaging techniques (e.g., ECoG, LFPs, single neuronal recordings) are currently unable to capture the entire complexity of expressive speech. Consequently, silent speech BCIs are limited in the vocabulary that can be decoded directly from the brain signals. The second avenue is to replace unnatural stimuli that require an artificial mapping with speech stimuli that do not. In such a system, the user would communicate simply by directing attention to the speech stimulus that matches his/her intent. Previous studies that explored this avenue required the speech stimuli to be designed [e.g., altered and broken up (Lopez-Gordo et al. 2012)] such that they elicit a particular and discriminable evoked response. Such evoked responses can be readily detected in scalp-recorded electroencephalography (EEG) to identify the attended speech stimulus. However, such altered speech stimuli are difficult to understand, which makes such a BCI system difficult to use. More importantly, this approach does not scale well beyond two simultaneously presented speech stimuli.

Recent studies suggest that the envelope of attended speech is directly tracked by electrocorticographic (ECoG) signals in the gamma band (i.e., 70–170 Hz) (Martin et al. 2014; Potes et al. 2012, 2014; Pasley et al. 2012; Kubanek et al. 2013), effectively removing the need to 'alter' the speech stimuli. Further evidence shows that this approach can identify auditory attention to one speaker in a mixture of speakers, i.e., a 'cocktail-party' situation (Zion et al. 2013).

However, BCI systems that use this physiological mechanism for communication purposes have not been described yet. In this study, we explore this possibility by implementing a BCI2000-based real-time system that uses ECoG signals to identify the attended speaker.

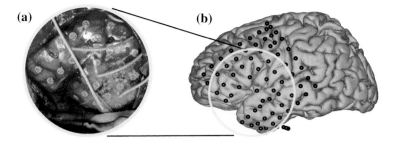

Fig. 1 *Implant* The subject had 72 subdural electrodes (1 grid and 3 strips in different configurations) implanted over left frontal, parietal, and temporal regions. **a** Photograph of the craniotomy and the implanted grids in this subject. **b** Cortical model of the subject's brain, showing an 8 × 8 grid over frontal/parietal cortex, and two strips

Methods

Human Subject

The subject in this study was a 49 year old left handed woman with intractable epilepsy who underwent temporary placement of subdural electrode arrays (see Fig. 1a) to localize seizure foci prior to surgical resection. A neuropsychological evaluation (Wechsler 1997) revealed normal cognitive function and hearing (full scale IQ = 97, verbal IQ = 91, performance IQ = 99) and a pre-operative Wada test (Wada and Rasmussen 1960) determined left hemispheric language dominance.

The subject had a total of 72 subdural electrode contacts (one 8 × 8 64-contact grid with 3 contacts removed, two strips in 1 × 4 configuration, and one strip in 1 × 3 configuration). The grid and strips were placed over the left hemisphere in frontal, parietal and temporal regions (see Fig. 1b for details). The implants consisted of flat electrodes with an exposed diameter of 2.3 mm and an inter-electrode distance of 1 cm, and were implanted for 1 week. Grid placement and duration of ECoG monitoring were based solely on the requirements of the clinical evaluation, without any consideration of this study. The subject provided informed consent, and the study was approved by the Institutional Review Board of Albany Medical College.

We used post-operative radiographs (anterior-posterior and lateral) and computed tomography (CT) scans to verify the cortical location of the electrodes. We then used Curry software (Neuroscan Inc, El Paso, TX) to create subject-specific 3D cortical brain models from high-resolution pre-operative magnetic resonance imaging (MRI) scans. We co-registered the MRIs by means of the post-operative CT and extracted the electrode coordinates according to the Talairach Atlas (Talairach and Tournoux 1988). These electrode coordinates are depicted on Talairach template brain in Fig. 1b.

Data Collection

We recorded ECoG from the implanted electrodes using a g.HIamp amplifier/digitizer system (g.tec, Graz, Austria) and the BCI software platform BCI2000 (Schalk et al. 2004; Mellinger et al. 2007; Schalk and Mellinger 2010), which sampled the data at 1200 Hz. Simultaneous clinical monitoring was implemented using a connector that split the cables coming from the patient into one set that was connected to the clinical monitoring system and another set that was connected to the g.HIamp devices. This ensured that clinical data collection was not compromised at any time. Two electrocorticographically silent electrodes (i.e., locations that were not identified as eloquent cortex by electrocortical stimulation mapping) served as electrical ground and reference, respectively.

Stimuli and Task

The subject's task was to selectively attend to one of two simultaneously presented speakers in a cocktail party situation (see Fig. 2a). The two speakers were John F. Kennedy and Barack Obama, each delivering his presidential inauguration address. Both speeches were similar in their linguistic features, but were uncorrelated in their sound intensities ($r = -0.02$, $p = 0.9$). To create a cocktail party situation, we mixed the two (monaural) speeches into a binaural presentation in which the stream presented to each ear contained 80 %:20 % of the volume of one speaker and 80 %:20 % of the other, respectively. This allowed us to manipulate the aural location of each speaker throughout the task. For the binaural presentation, we used in-ear monitoring earphones (AKP IP2, 12–23,500 Hz bandwidth) that isolated the subject from any ambient noise in the room.

To create a trial structure, we broke these combined streams into segments of 15–25 s in length, which resulted in a total of 10 segments of 187 s combined length. In the course of the experiment, we presented each segment four times to counter-balance the aural location (i.e., left and right) and the identity (i.e., JFK and Obama) of the attended speaker. Thus, over these four trials, the subjects had to attend to each of the two speakers at each of the two aural locations. This resulted in a total of 40 trials (i.e., 10 segments, each presented 4 times).

At the beginning of each trial, an auditory cue indicated the aural location (i.e., left or right) to which the subject should attend. Throughout the trial, a visual stimulus complemented the initial auditory cue to indicate the identity of the aural location (e.g., 'JFK in LEFT ear'). Each trial consisted of a 4 s cue, a 15–25 s stimulus and a 5 s inter-stimulus period. The total length of these 40 trials was 12.5 min. The subject performed these 40 trials in 5 blocks of 8 trials each, with a 3 min break between each block.

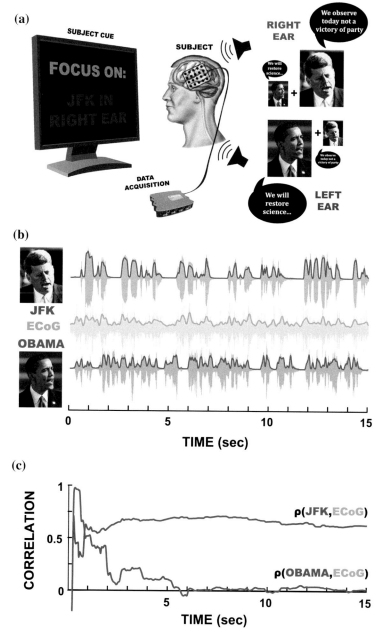

Fig. 2 *Experimental setup and methods* **a** Subjects selectively directed auditory attention to one of two simultaneously presented speakers. **b** We extracted the envelope of ECoG signals in the high gamma band, as well as the envelopes of the attended and unattended speech stimuli (i.e., JFK and Obama). **c** The correlation between the envelopes of the ECoG gamma band and the attended speech stimulus, accumulated over time, is markedly larger than the accumulated correlation between the envelopes of the ECoG gamma band and the unattended speech stimulus

Offline Analysis

In the offline analysis, we characterized the relationship between the neural response (i.e., the ECoG signals) and the attended and unattended speech streams, as shown in Fig. 2b. In particular, we were interested in two parameters of this neural response. The first parameter was the delay between the audio stream and resulting cortical processing, i.e., the time from presentation of the audio stream to the observation of the cortical change. The second parameter was the cortical location that was most selective to the attended speech stream. These two parameters are the only two parameters that were later needed to configure the online BCI system.

To determine these two parameters, we extracted the high gamma band envelope at each cortical location and the envelopes of the covertly attended and unattended speech (i.e., JFK and Obama). We then correlated the high gamma band envelope at each cortical location, once with the attended and once with the unattended speech envelope. This resulted in two Spearman's r-values for each cortical location. An example of this is shown in Fig. 2c. To determine the delay between the audio stream and resulting cortical processing, we measured the neural tracking of the sound intensity across different delays from 0 to 250 ms to identify the deal with the highest r-value.

Signal Processing

We first pre-processed the ECoG signals from the 72 channels to remove external noise. To do this, we high-pass filtered the signals at 0.5 Hz and re-referenced them to a common average reference that we composed from only those channels for which the 60 Hz line noise was within 1.5 standard deviations of the average.

Next, we extracted the signal envelope in the high gamma band using these pre-processed ECoG signals. For this, we applied an 18th order 70–170 Hz Butterworth filter and then extracted the envelope of the filtered signals using the Hilbert transform. Finally, we low-pass filtered the resulting signal envelope at 6 Hz (anti-aliasing) and downsampled the result to 120 Hz.

For each attended und unattended auditory stream, we extracted the time course of the sound intensity, i.e., the envelope of the signal waveform in the speech band. To do this, we applied a 80–6000 Hz Butterworth filter to each audio signal, and then extracted the envelope of the filtered signals using the Hilbert transform. Finally, we low-pass filtered the speech envelopes at 6 Hz and downsampled them to 120 Hz.

Feature Extraction

We extracted features that reflect the neural tracking of the attended and unattended speech stream. We defined neural tracking of speech as the correlation between the gamma envelope (of a given cortical location) and the speech envelope. We calculated this correlation separately for the attended and unattended speech, thereby obtaining two sets of r-values labeled 'attended' and 'unattended,' respectively.

Selection of Cortical Delay and Location

We expected a delay between the audio presentation and resulting cortical processing, i.e., the time from presentation of the audio stimuli to the observation of the cortical change. To account for this delay, we measured the neural tracking of the attended speech stream across different delays (0–250 ms, see Fig. 3) and across all channels. Next, we determined the cortical location that was most selective of the attended speech stream. For this, we selected the cortical location that showed the largest difference between the 'attended' and 'unattended' r-values. Based on these results, we selected a delay of 150 ms and a cortical location over superior temporal gyrus (STG). We corrected for this delay by shifting the speech envelopes relative to the ECoG envelopes prior to calculating the correlation values.

Classification

In our approach, we assumed that the extracted features, i.e., the two r-values of the selected cortical location, were directly predictive of the 'attended' conversation. In other words, for the selected cortical location, if the 'attended' r-value was larger than the 'unattended' r-value, the the trial was classified correctly. To determine the performance as a function of the length of attention, we applied our feature extraction and classification procedure to data segments from 0.1 to 15 s in length.

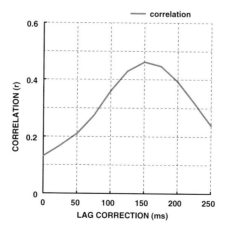

Fig. 3 *Lag between speech presentation and neural response* This figure shows the correlation between neural response and the attended speech (*green*) for the most selective cortical location, across corrected lags between 0 and 250 ms. This correlation peaks at 150 ms

Real-Time System Verification

In the real-time verification, we evaluated the system performance on the data recorded during the first stage of this study. We configured this system with parameters (i.e., cortical location and delay) determined in the previously detailed offline analysis.

Real-Time System Architecture

We used the BCI software platform BCI2000 (Schalk et al. 2004; Mellinger et al. 2007; Schalk and Mellinger 2010) to implement an auditory attention based BCI. For this, we expanded BCI2000 with the capability to process auditory signals in real time. In detail, we implemented a signal acquisition for audio devices (e.g., a microphone) or pre-recorded files that is synchronized with the acquisition from the neural signals. Further, we implemented a signal correlation filter. For our evaluation, the two (monaural) speeches served as the audio input to the auditory attention based BCI (see Fig. 4).

   In this system, BCI2000 filters the audio signals between 80–6000 Hz and the ECoG signals between 70–170 Hz. Next, a BCI2000 filter extracts the envelopes, decimates them to a common sampling rate of 200 Hz and adjusts their timing for the cortical delay. A signal correlation filter then calculates the correlation values, i.e., the correlation between the two (monaural) speeches and the selected neural envelope, to determine to which speaker the user directs his/her attention. Finally, the feedback augmentation filter increases the volume of the attended speaker and decreases the volume of the unattended speaker to provide feedback to the subject. This processing steps are updated every 50 ms to provide feedback in real-time.

Results

Neural Correlates of Attended and Unattended Speech

First, we were interested in visualizing the cortical areas that track the 'attended' and 'unattended' conversations. The results in Fig. 5 show the neural tracking of the 'attended' and 'unattended' speech in the form of an activation index. For each cortical location, this activation index expresses the negative logarithm of the p-value ($-\log(p)$) of the correlation between the high gamma ECoG envelope and the attended or unattended speech envelope. The neural tracking is focused predominantly on areas on or around superior (STG) and middle temporal gyrus (MTG).

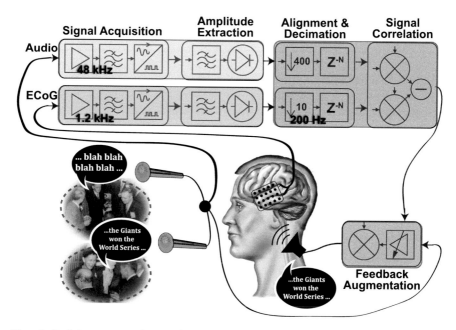

Fig. 4 *Real-time system design* The auditory attention BCI is based on BCI2000 and simultaneously acquires and processes **audio** and **ECoG** signals. The audio signals from multiple conversations are sampled at 48 kHz and acquired from a low-latency USB audio-amplifier (Tascam US-122MKII). The ECoG signals from the surface of the brain are sampled at 1200 Hz and acquired from a 256-channel bio-signal amplifier (g.HIamp, g.tec Austria). In the next step, the signals are band-pass filtered (80–6000 Hz for audio, 70–170 Hz for ECoG) and their envelope is extracted. The resulting signal envelopes are decimated to a common sampling rate of 200 Hz and adjusted for timing differences. One channel of the decimated ECoG signal envelope is then selected and correlated with each of the decimated audio signal envelopes. As the human subject perceives the mixture of conversations through ear-phones, the auditory attention BCI then can provide feedback by modifying the volume of the presented mixture of conversations to enhance the volume of the attended and attenuate the volume of the unattended conversation

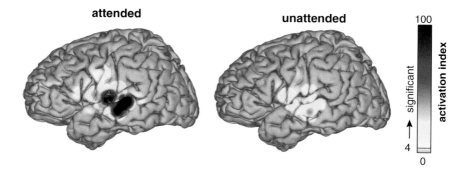

Fig. 5 *Neural tracking of attended* (*left*) *and unattended* (*right*) *speech* The tracking of the attended speech is both stronger and more widely distributed than the tracking of the unattended speech. In addition, there is only a marginal difference in spatial distribution between attended and unattended stimuli

Relationship Between Segment Length and Classification Accuracy

Next, we were interested in determining the duration of attention that is needed to infer the 'attended' speech. For this, we examine the relationship between the segment length and classification accuracy. The results in Fig. 6 show the classification accuracies for variable segment lengths (0.1–15 s). In this graph, the accuracy improvements level off after 5 s, at 80–90 % accuracy.

Interface to the Investigator

Finally, we evaluated the real-time system performance that the determined parameters (i.e., the cortical location and delay) yield on the data recorded during the first stage of this study. The screenshot in Fig. 7 shows interface to the investigator. The interface presents the decimated and aligned ECoG and audio envelopes, their correlation with each other, and the inferred attention. The content of this interface is updated 20 times per second.

Discussion

We show the first real-time implementation of an auditory attention BCI that uses ECoG signals and natural speech stimuli. The configuration of this system requires only two parameters: the cortical location and the delay between the audio presentation and the cortical processing. Our results can guide the selection of these parameters. For example, our results indicate that the underlying physiological mechanism is primarily focused on the temporal lobe, specifically the STG and MTG areas. Further, the neural tracking of attended speech is stronger and more

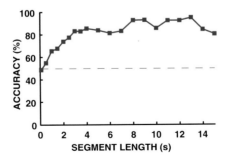

Fig. 6 *Accuracy for different segment lengths* The classification accuracy generally increases with segment length. The red horizontal dashed line indicates chance accuracy

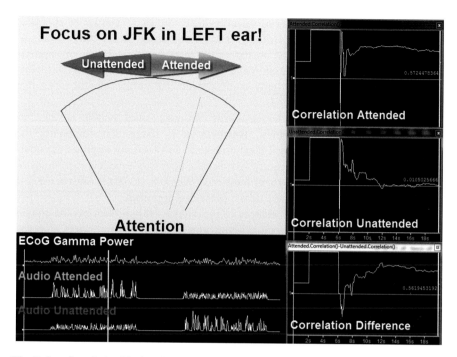

Fig. 7 *Interface design* The interface to the investigator presents multiple panels. The bottom left panel presents the decimated and aligned ECoG and audio envelopes. The panels on the right show the correlation between the ECoG and the attended (top), ECoG and unattended (middle) and the difference between the two correlation values (bottom). The panel on the top left shows this correlation difference in form of an analogue instrument where the pointer (i.e., the needle) indicates the direction of attention. In this experiment, the subject was cued to attend to a particular speaker annotated by "Attended" in this panel

widely distributed than that of unattended speech. This confirms results from a previous ECoG study that investigated auditory attention (Zion et al. 2013). Further, our study shows that the cortical delay between the audio presentation and the cortical processing is in the range of ~ 150 ms.

The presented results indicate that such system could support BCI communication. While being invasive, it may be justified for those affected by severe neuro-degenerative diseases (e.g., late-stage ALS, locked-in syndrome) who have lost all muscular control and therefore cannot use conventional assistive devices or BCIs that depend on visual stimulation or feedback. Most importantly, the results suggest that sufficient communication performance (>70 %, Kübler et al. 2001) could be achieved with a single electrode placed over STG or MTG. This finding is important, because placement of ECoG grids as used in this study requires a large craniotomy. In contrast, a single electrode could be placed through a burr hole (Leuthardt et al. 2009). Furthermore, the electrodes in this study were placed subdurally (i.e., the electrodes are placed underneath the dura). Penetration of the dura increases the risk of bacterial infection (Davson 1976; Hamer et al. 2002; Fountas and Smith 2007; Van Gompel et al. 2008; Wong et al. 2009). Epidural electrodes (i.e., electrodes placed on top of the dura) provide signals of approximately comparable fidelity (Torres Valderrama et al. 2010; Bundy et al. 2014). A single electrode placed epidurally could reduce risk, which should make this approach more clinically practical.

In this study, we focused on demonstrating that one cortical location is sufficient for providing BCI communication. However, it is likely that combining the information from multiple cortical locations could substantially improve the communication performance. Thus, recent advances in clinically practical recordings of ECoG signals from multiple cortical locations (Sillay et al. 2013; Stieglitz 2014) could improve the clinical efficacy of the presented approach.

In comparison to many other auditory BCIs, the present approach has the unique advantage of using natural speech without any alteration. This aspect may be particularly relevant for those who are already at a stage where learning how to use a BCI has become difficult.

Conclusion

In summary, our study demonstrates the function of an auditory attention BCI that uses ECoG signals and natural speech stimuli. The implementation of this system within BCI2000 lays the groundwork for future studies that investigate the clinical efficacy of this system. Once clinically evaluated, such a system could provide communication without depending on other sensory modalities or a mapping between the stimulus and the communication intent. In the near future, this could substantially benefit people affected by severe motor disabilities that cannot use conventional assistive devices or BCIs that require some residual motor control, including eye movement.

Acknowledgements This work was supported by the NIH (EB006356 (GS), EB00856 (GS) and EB018783 (GS)), the US Army Research Office (W911NF-07-1-0415 (GS), W911NF-08-1-0216 (GS) and W911NF-14-1-0440 (GS)) and Fondazione Neurone.

References

A. Belitski, J. Farquhar, P. Desain, J. Neural Eng. **8**(2), 025022 (2011). doi:10.1088/1741-2560/8/2/025022

A.M. Brouwer, J.B. van Erp, Front Neurosci **4**, 19 (2010). doi:10.3389/fnins.2010.00019

P. Brunner, G. Schalk, Clin. Neurophysiol. (2010). doi:10.1016/j.clinph.2010.11.014

P. Brunner, S. Joshi, S. Briskin, J.R. Wolpaw, H. Bischof, G. Schalk, J. Neural Eng. **7**(5), 056013 (2010). doi:10.1088/1741-2560/7/5/056013

D.T. Bundy, E. Zellmer, C.M. Gaona, M. Sharma, N. Szrama, C. Hacker, Z.V. Freudenburg, A. Daitch, D.W. Moran, E.C. Leuthardt, J. Neural Eng. **11**(1), 016006 (2014). doi:10.1088/1741-2560/11/1/016006

H. Davson, J. Physiol. **255**(1), 1 (1976)

K.N. Fountas, J.R. Smith, Stereotact. Funct. Neurosurg. **85**(6), 264 (2007). doi:10.1159/000107358

A. Furdea, S. Halder, D.J. Krusienski, D. Bross, F. Nijboer, N. Birbaumer, A. Kübler, Psychophysiology **46**(3), 617 (2009). doi:10.1111/j.1469-8986.2008.00783.x

S. Halder, M. Rea, R. Andreoni, F. Nijboer, E.M. Hammer, S.C. Kleih, N. Birbaumer, A. Kübler, Clin. Neurophysiol. **121**(4), 516 (2010). doi:10.1016/j.clinph.2009.11.087

H.M. Hamer, H.H. Morris, E.J. Mascha, M.T. Karafa, W.E. Bingaman, M.D. Bej, R.C. Burgess, D.S. Dinner, N.R. Foldvary, J.F. Hahn, P. Kotagal, I. Najm, E. Wyllie, H.O. Lüders, Neurology **58**(1), 97 (2002)

D.S. Klobassa, T.M. Vaughan, P. Brunner, N.E. Schwartz, J.R. Wolpaw, C. Neuper, E.W. Sellers, Clin. Neurophysiol. **120**(7), 1252 (2009)

J. Kubanek, P. Brunner, A. Gunduz, D. Poeppel, G. Schalk, PLoS One **8**(1), e53398 (2013). doi:10.1371/journal.pone.0053398

A. Kübler, B. Kotchoubey, J. Kaiser, J.R. Wolpaw, N. Birbaumer, Psychol. Bull. **127**(3), 358 (2001)

E.C. Leuthardt, Z. Freudenberg, D. Bundy, J. Roland, Neurosurg. Focus **27**(1), E10 (2009). doi:10.3171/2009.4.FOCUS0980

E.C. Leuthardt, C. Gaona, M. Sharma, N. Szrama, J. Roland, Z. Freudenberg, J. Solis, J. Breshears, G. Schalk, J. Neural Eng. **8**(3), 036004 (2011). doi:10.1088/1741-2560/8/3/036004

M.A. Lopez-Gordo, E. Fernandez, S. Romero, F. Pelayo, A. Prieto, J. Neural Eng. **9**(3), 036013 (2012). doi:10.1088/1741-2560/9/3/036013

F. Lotte, J.S. Brumberg, P. Brunner, A. Gunduz, A.L. Ritaccio, C. Guan, G. Schalk, Front Hum Neurosci **9**, 97 (2015). doi:10.3389/fnhum.2015.00097

S. Martin, P. Brunner, C. Holdgraf, H.J. Heinze, N.E. Crone, J. Rieger, G. Schalk, R.T. Knight, B. Pasley, Front Neuroeng **7**(14) (2014). doi:10.3389/fneng.2014.00014

J. Mellinger, G. Schalk, in *Toward Brain-Computer Interfacing*, ed. by G. Dornhege, J. del R. Millan, T. Hinterberger, D. McFarland, K. Müller (MIT Press, Cambridge, 2007), pp. 359–367

B.N. Pasley, S.V. David, N. Mesgarani, A. Flinker, S.A. Shamma, N.E. Crone, R.T. Knight, E.F. Chang, PLoS Biol. **10**(1), e1001251 (2012). doi:10.1371/journal.pbio.1001251

X. Pei, D.L. Barbour, E.C. Leuthardt, G. Schalk, J. Neural Eng. **8**(4), 046028 (2011). doi:10.1088/1741-2560/8/4/046028

X. Pei, J. Hill, G. Schalk, IEEE pulse **3**(1), 43 (2012). doi:10.1109/MPUL.2011.2175637

C. Potes, A. Gunduz, P. Brunner, G. Schalk, NeuroImage **61**(4), 841 (2012). doi:10.1016/j.neuroimage.2012.04.022

C. Potes, P. Brunner, A. Gunduz, R.T. Knight, G. Schalk, NeuroImage **97**, 188 (2014). doi:10.1016/j.neuroimage.2014.04.045

A. Riccio, D. Mattia, L. Simione, M. Olivetti, F. Cincotti, J. Neural Eng. **9**(4), 045001 (2012). doi:10.1088/1741-2560/9/4/045001

G. Schalk, J. Mellinger, *A Practical Guide to Brain-Computer Interfacing with BCI2000*, 1st edn. (Springer, London, 2010)

G. Schalk, D.J. McFarland, T. Hinterberger, N. Birbaumer, J.R. Wolpaw, IEEE Trans. Biomed. Eng. **51**(6), 1034 (2004)

M. Schreuder, B. Blankertz, M. Tangermann, PLoS One **5**(4) (2010). doi:10.1371/journal.pone.0009813

K.A. Sillay, P. Rutecki, K. Cicora, G. Worrell, J. Drazkowski, J.J. Shih, A.D. Sharan, M. J. Morrell, J. Williams, B. Wingeier, Brain Stimulation **6**(5), 718 (2013)

T. Stieglitz, **7**, 9 (2014). doi:10.1007/978-3-319-08072-7_3

J. Talairach, P. Tournoux, *Co-Planar Sterotaxic Atlas of the Human Brain* (Thieme Medical Publishers Inc, New York, 1988)

A. Torres Valderrama, R. Oostenveld, M.J. Vansteensel, G.M. Huiskamp, N.F. Ramsey, J. Neurosci. Methods **187**(2), 270 (2010). doi:10.1016/j.jneumeth.2010.01.019

M. van der Waal, M. Severens, J. Geuze, P. Desain, J. Neural Eng. **9**(4), 045002 (2012). doi:10.1088/1741-2560/9/4/045002

J.J. Van Gompel, G.A. Worrell, M.L. Bell, T.A. Patrick, G.D. Cascino, C. Raffel, W.R. Marsh, F. B. Meyer, Neurosurgery **63**(3), 498 (2008). doi:10.1227/01.NEU.0000324996.37228.F8

J. Wada, T. Rasmussen, J. Neurosurg. **17**, 266 (1960)

D. Wechsler, *Weschsler Adult Intelligence Scale-III* (The Psychological Corporation, San Antonio, 1997)

J.R. Wolpaw, N. Birbaumer, D.J. McFarland, G. Pfurtscheller, T.M. Vaughan, Clin. Neurophysiol. **113**(6), 767 (2002). doi:10.1016/S1388-2457(02)00057-3

C.H. Wong, J. Birkett, K. Byth, M. Dexter, E. Somerville, D. Gill, R. Chaseling, M. Fearnside, A. Bleasel, Acta Neurochir. (Wien) **151**(1), 37 (2009). doi:10.1007/s00701-008-0171-7

E.M. Zion Golumbic, N. Ding, S. Bickel, P. Lakatos, C.A. Schevon, G.M. McKhann, R.R. Goodman, R. Emerson, A.D. Mehta, J.Z. Simon, D. Poeppel, C.E. Schroeder, Neuron **77**(5), 980 (2013). doi:10.1016/j.neuron.2012.12.037

Neurofeedback Training with a Motor Imagery-Based BCI Improves Neurocognitive Functions in Elderly People

J. Gomez-Pilar, R. Corralejo, D. Álvarez and R. Hornero

Abstract In recent years, brain-computer interfaces (BCIs) have become not only a tool to provide communication and control for people with disabilities, but also a way to rehabilitate some motor or cognitive functions. Brain plasticity can help restore normal brain functions by inducing brain activity. In fact, voluntary event-related desynchronization (ERD) in upper alpha and beta electroencephalo-gram (EEG) activity bands is associated with different neurocognitive functions. In this regard, neurofeedback training (NFT) has shown to be a suitable way to control one's own brain activity. Furthermore, new evidence in recent studies showed NFT could lead to microstructural changes in white and grey matter. In our novel study, NFT qualities were applied to aging-related effects. We hypothesized that a NFT by means of motor imagery-based BCI (MI-BCI) could affect different cognitive functions in elderly people. To assess the effectiveness of this application, we studied 63 subjects, all above 60 years old. The subjects were divided into a control group (32 subjects) and a NFT group (31 subjects). To validate the effectiveness of the NFT using MI-BCI, variations in the scores of neuropsychological tests (Luria tests) were measured and analyzed. Results showed significant improvements ($p < 0.05$) in the NFT group, after only five NFT sessions, in four cognitive functions: visuospatial; oral language; memory; and intellectual. These results further support the association between NFT and the enhancement of cognitive performance. Findings showed the potential usefulness of NFT using MI-BCI. Therefore, this approach could lead to new means to help elderly people.

J. Gomez-Pilar (✉) · R. Corralejo · D. Álvarez · R. Hornero
Biomedical Engineering Group, E.T.S.I. Telecomunicación,
Universidad de Valladolid, Paseo Belén 15, 47011 Valladolid, Spain
e-mail: javier.gomez@gib.tel.uva.es

R. Corralejo
e-mail: rebeca.corralejo@gib.tel.uva.es

D. Álvarez
e-mail: dalvgon@gmail.com

R. Hornero
e-mail: roberto.hornero@tel.uva.es

© The Author(s) 2015
C. Guger et al. (eds.), *Brain-Computer Interface Research*,
SpringerBriefs in Electrical and Computer Engineering,
DOI 10.1007/978-3-319-25190-5_5

43

Keywords Neurofeedback   ·   Brain-Computer   interface   (BCI)   ·
Electroencephalogram (EEG) · Luria adult neuropsychological diagnosis (AND) ·
Elderly people

Introduction

Elderly people undergo numerous changes that imply poorer cognitive performance
than young adults. Some cognitive tasks that are impaired due to the aging are
visuospatial perception, memory and attention (Craik and Salthouse 2011).
Although some older adults perform cognitive tasks as well as young people,
cognitive deficit is one of the most prominent concerns in the elderly (Christensen
1979). Furthermore, according to the United Nations (2013), one in every 3 persons
in developed countries will be 60 years or older in 2050. Therefore, it seems clear
that the study of applications aimed at helping the elderly is of paramount
importance.

Neurofeedback training (NFT) is a promising approach to restore some cognitive
functions. By encouraging healthy brain activity, brain plasticity can help restore
normal brain functions (Daly and Wolpaw 2008). Many studies indicated the
effectiveness of NFT in the treatment of several mental disorders: attention deficit
hyperactivity disorder (Arns et al. 2014); epilepsy (Sterman and Egner 2006);
autism (Coben et al. 2010); and traumatic brain injury (Thornton and Carmody
2008), among others. Furthermore, there are some reports indicating that NFT
might be used to increase cognitive performance (Vernon et al. 2003). Hence, it
seems reasonable to explore NFT to restore cognitive performance that has been
reduced in elderly persons. In this regard, there are studies that focus on neuropsychological changes due to NFT in healthy elderly people (Staufenbiel et al.
2014; Wang and Hsieh 2013; Angelakis et al. 2006). The main problem is the
extended controversy about the effects on cognitive performance or the transference
of the feedback beyond the training sessions. While some studies have found no
cognitive effects due to NFT (Staufenbiel et al. 2014), others found an increase in
cognitive processing speed and executive functions but not in memory (Angelakis
et al. 2006). Therefore, further analyses of NFT effects in elderly are needed.

The main novelty in our study is the use of a brain-computer interface (BCI) as a
way to provide NFT. Since NFT can be a hard task for real users, we felt that the
design of a NFT protocol via a BCI merits extra attention. This could affect the
number of sessions needed to yield changes in cognitive functions. Therefore, in
our study, participants performed the NFT with a BCI focused on age-related
impairments. The designed NFT consists of five different games controlled by a
motor imagery-based BCI (MI-BCI).

We hypothesized that repetitive stimulation of endogenous brain activity in
certain cortical areas improves brain plasticity. This assumption is based on some
evidences of neuroplastic changes occurring after NFT (Ros et al. 2010) joined by

microstructural changes in white and grey matter (Ghaziri et al. 2013). Therefore, it stands to reason that continued training of certain brain regions that have lost plasticity helps enhance cognitive performance. The aim of this study was to assess cognitive changes produced by NFT through neuropsychological tests that evaluate the main cognitive tasks reduced due to the aging.

In this study, we addressed three research questions: (i) can our designed NFT protocol enhance cognitive performance in elderly people?; (ii) are these cognitive changes statistically significant compared to a control group?; and (iii) can the use of BCI reduce the number of sessions needed to find significant changes compared to a typical NFT protocol?

Methods

Population Studied

A total of 63 subjects participated in the experiment. Since this study was focused in elderly people, all participants were older than 60. All subjects were generally healthy, with a similar educational level. None of them had any BCI previous experience (BCI-naives). The population was divided into a control group (32 subjects) and a NFT group (31 subjects). The control group was composed by 23 females and 9 males (mean age = 68.0 ± 5.6 years, range = 60–80), while the NFT group consisted of 18 females and 13 males (mean age = 68.3 ± 4.3 years, range = 60–81). Nonsignificant differences were observed in age or gender ($p > 0.05$, Mann-Whitney U-test) between both groups.

EEG Recordings

EEG was recorded using 8 active electrodes (F3, F4, T7, C3, Cz, C4, T8 and Pz) placed in an elastic cap, according to the international 10–20 system (Jasper 1958). Data were referenced to a common reference placed on the ear lobe. The ground electrode was located at channel AFz. EEG data were filtered online with a bandwidth of 0.1–60 Hz. A notch filter was also used to remove the power line frequency interference (50 Hz in Spain). Signals were amplified by a g.USBamp amplifier (Guger Technologies OG, Graz, Austria) and digitally stored at a sampling rate of 256 Hz. A first neighbor's Laplacian filter over C3, Cz and C4 was used for preprocessing to help provide feedback to the users. Specifically, spectral bands of 3 Hz centered on 12, 18 and 21 Hz were used to train upper alpha and beta frequency bands.

We developed a novel BCI tool aimed at performing NFT tasks. The BCI2000 general purpose system (Schalk et al. 2004) was used for real time EEG recordings and signal processing in the NFT group.

Design of the Experiment

The experiment consisted of four steps.

1. **Pre-test step** Control and NFT groups performed a neuropsychological test called the Luria Adult Neuropsychological Diagnosis (AND), which is typically used with healthy subjects (Christensen 1979). This pre-test provides the basis to understand neuropsychological characteristics of all subjects before the following steps. Luria-AND test is a more comprehensive way of measuring cognitive abilities than the tests usually applied in previous similar studies (e.g. mental rotation tasks).
2. **NFT step** Only the NFT group performed different NFT tasks, across 5 sessions (once per week), based on our MI-BCI application that provided real time feedback. Each session lasted about 1 h. This step is one of the main novelties of this study. Previous studies used sensorimotor rhythms for performing NFT (Pineda et al. 2003; 2014). In these works, certain frequency bands were trained. However, MI-BCI is a relatively novel alternative in order to assess cognitive changes in elderly users. Thanks to BCI, the NFT protocol is relatively simple to perform. BCIs also facilitate friendly interface design, so that participants are more predisposed to perform the training.
3. **Post-test step** All subjects (control and NFT groups) performed the Luria-AND test again about 2 months after pre-test. Thus, potential changes in different neuropsychological functions can be measured with this post-test.
4. **Offline analysis step** An offline analysis was carried out using the Luria-AND pre- and post-scores to verify the influence of NFT over the different cognitive functions.

Motor Imagery-Based BCI Application

NFT was carried out via a BCI to get a friendly environment for the users. Therefore, the participants performed the tasks with constant feedback to learn to control upper alpha and beta frequency bands. This approach was intended to increase the motivation of users to make the training more effective. The MI-BCI users' mental task was to imagine repetitive movements of the left or right hand. If this task was properly performed, alpha and beta rhythms were activated. Hence, different cortical regions were stimulated, affecting several cognitive functions.

For this purpose, five NFT tasks (T1, T2, T3, T4 and T5) were designed and developed. These tasks were:

- **T1** The aim of the first task was to learn how to imagine hand movements. For this purpose, the application showed the user a door or a window, randomly. When the door is displayed, the user had to imagine right hand movements. When a window was shows, the user had to imagine left hand movements. If the exercise is done properly, the application shows feedback to the user so that the door or window is opened. Therefore, the user can understand when the task was correctly performed, or when the imagery strategy must be changed. Figure 1 shows screenshots at different points in the execution of this task.
- **T2** This task was designed to move a cursor horizontally via motor imagery exercises. Users were asked to move an object on a monitor (a person, pair of trousers or fish) by imagining hand movements. They had to move this object to a target (a house, a wardrobe or a fridge, respectively) on the right or left of the screen. When the target was reached, the object disappeared by entering the target through a door. However, if the object did not reach the target, the door remained closed. Users received continuous feedback from the application, since the object's position was continually updated. Figure 2 shows screenshots of this exercise.

Fig. 1 Screenshots of the interface during T1. The two *left panels* show the window and the door before the exercise begins. The two *right panels* show the window and door after being opened after correct task performance

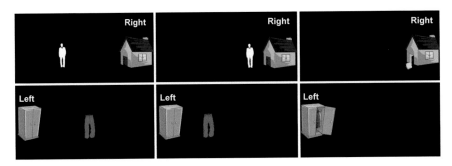

Fig. 2 Interface screenshots during T2. The *top panels* represent feedback corresponding to *right hand* motor imagery. The *bottom panels* present *left hand* imagery feedback

- **T3** This exercise consisted of choosing the correct direction and then moving an object on the monitor. To choose the correct direction, users had to solve simple and logical relations between items: fish or meat, which are related to a fridge, and shirt or trousers, which are related to a wardrobe. Participants received continuous feedback from the screen via object movement and the position of the door. Figure 3 shows screenshots of this task.
- **T4** This exercise is presented as a game. Users had to control a virtual object, represented by a person, to overcome different obstacles (animals, rocks or vehicles). Continuous feedback was presented as before. The velocity of the person can be modified increasing or decreasing game difficulty. Every time an obstacle is avoided, a message with positive reinforcement appears in the screen. Figure 4 shows screenshots of this task in different stages of the game.
- **T5** This task, the most complex one, was composed of three stages: (i) first, two images were displayed on the screen for 3 s, (ii) then, they disappeared and two images were shown on the right and the left of the screen, only one of which matched initial images, and (iii) finally, the user had to identify which one of these images appeared at the beginning of the trial and move the cursor towards it. Hence, T5 combines hand motor imagery tasks with working memory exercises. Figure 5 shows screenshots of each of the three stages of this task.

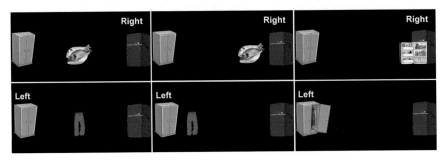

Fig. 3 Screenshots of the interface during T3. The *top panels* represent feedback during *right hand* motor imagery. The *bottom panels* reflect *left hand* motor imagery feedback

Fig. 4 Screenshots of the interface during T4. Users must overcome obstacles using hand motor imagery

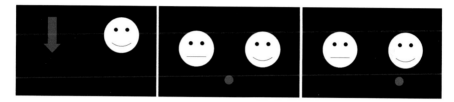

Fig. 5 Screenshots of the interface during T5. During the first stage (*left*), users had to memorize these two items. In the second stage (*center*), users had to identify the repeated item. At the last stage (*right*), users had to move the cursor (*red ball*) to this target. In this example, the user has correctly identified the repeated item (*a smiling face*) and successfully moved the cursor to this target

As Table 1 shows, the first sessions contained several trials of T1 and T2 to learn and to practice motor imagery strategies. In the following sessions, the complexity of the training was increased while users practiced complementary exercises such as logical relation (T3), attention (T4) or memory (T5). Although the main task during training was hand motor imagery, different brain regions were activated during NFT due to these complementary exercises. The number of trials in each session was adjusted so that each session lasted about 1 h.

Table 1 Trial distribution of each task among the five sessions

	Session 1	Session 2	Session 3	Session 4	Session 5
T1	150	60	0	0	0
T2	45	90	90	45	45
T3	0	60	48	32	0
T4	0	0	20	30	30
T5	0	0	0	30	75

Luria-AND Test

To provide an adequately thorough neuropsychological analysis of the study participants, both the control and NFT groups performed the Luria-AND test twice: (i) during the pre-test stage and (ii) during the post-test stage. It was thereby possible to evaluate the potential changes due to BCI-based NFT.

The Luria-AND test includes nine subtests that assess five different functions: visuospatial (visual perception and spatial orientation); oral language (receptive speech and expressive speech); memory (immediate memory and logical memory); intellectual (thematic draws and conceptual activity); and attention (attentional control) (Christensen 1979). This test is a useful way to assess cognitive features in healthy populations. Furthermore, as mentioned before, this test can also assess the main cognitive functions that may be impaired due to ageing—visuospatial perception, memory and attention.

Statistical Analysis

Cognitive changes between pre- and post-scores were measured. The Kolgomorov–Smirnov and Levene tests were applied to provide a descriptive analysis of these changes. After finding that the data did not meet parametric assumptions, non-parametric tests were chosen to evaluate our results. For intragroup analysis, the Wilcoxon signed–rank test ($p < 0.05$) was applied. The Mann–Whitney U–test (statistical significance $p < 0.05$) was used to assess the statistical differences in the scores of each neuropsychological feature between both groups. Hence, a total of five p-values were calculated for each cognitive feature:

- Comparison of pre-scores between control and NFT groups using the Mann–Whitney test.
- Comparison of post-scores between control and NFT groups using the Mann–Whitney test.
- Comparison of the variation of control and NFT groups between pre and post-scores using the Mann–Whitney test.
- Comparison of the pre and post-scores in the control group using the Wilcoxon test.
- Comparison of the pre and post-scores in the NFT group using the Wilcoxon test.

Thus, it was possible evaluate the statistical significance of the changes in neuropsychological scores that occurred.

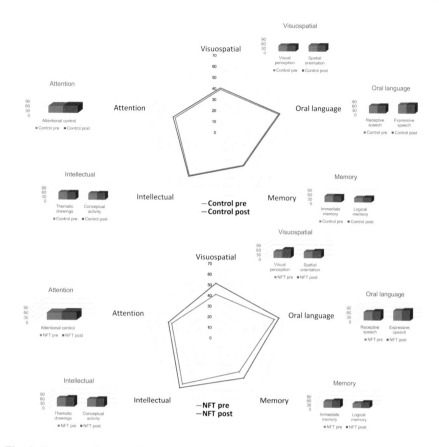

Fig. 6 Test scores for control (*top*) and NFT (*bottom*) groups. Bar diagrams show all cognitive functions assessed in this study. The NFT group shows a clear generalized increase in the post-scores

Results

Figure 6 shows changes in control and NFT groups. A visual inspection of these scores indicates that, overall, the scores of the control group showed no noteworthy changes between pre-test and post-test. However, the NFT group clearly exhibited changes between the pre- and post-test scores. Specifically, visuospatial and memory functions showed the most prominent variations.

Once changes in scores between pre- and post-tests were observed, we searched for statistically significant differences. Table 2 summarizes the statistical differences of Luria-AND scores for each cognitive function. Results reveal several relevant aspects:

Table 2 Statistical differences of Luria-AND tests scores for each cognitive function. Significant values ($p < 0.05$) are highlighted

Cognitive function	NFT pre versus control pre	NFT post versus control post	Δ Control group versus Δ NFT group	Δ Control group (pre vs. post)	NFT group (pre vs. post)
Visuospatial	0.504	**<0.05**	**<0.05**	0.535	**<0.05**
Oral language	0.423	**<0.05**	**<0.05**	0.661	**<0.05**
Memory	0.549	**<0.05**	**<0.05**	0.824	**<0.05**
Intelligence	0.072	**<0.05**	**<0.05**	0.726	**<0.05**
Attention	0.363	0.166	0.986	0.662	0.211
SNV	0.158	**<0.05**	**<0.05**	0.489	**<0.05**

- The analysis of intergroup pre-scores suggests that both groups (NFT and control groups) presented similar distributions for each cognitive function before NFT.
- Regarding the intergroup post-scores, there are significant differences in the score distribution between both groups for four cognitive functions: visuospatial; oral language; memory and intellectual.
- Comparisons of control and NFT groups assessing pre and post-scores show statistical differences in the same four functions, even after only five sessions.
- The same comparison did not yield statistically significant differences in the control group.
- The intragroup analysis for the NFT group again reveals significant changes in the same four functions after the BCI-based NFT program.

Finally, by normalizing the test scores, data can be clustered into a single neurocognitive value (*SNV*) for all tests. Hence, the neurocognitive scores of *SNV* for both groups were compared, with *P*-values shown in Table 2. These results support a generalized increase in the neurocognitive scores by means of BCI-based NFT.

Discussion

This study introduced a novel NFT strategy based on BCI using hand motor imagery to train specific frequency bands (upper alpha and beta) of the EEG. We focused on NFT effects in elderly people (over 60 years old). We found no significant differences in the comparison between the control group and the NFT group based on Luria-AND values during the pre-test. This means that the population had similar neuropsychological scores before starting NFT. However, after NFT, statistical significant differences in post-scores were found between control and NFT groups across four cognitive functions: visuospatial; oral language; memory and intellectual. The intergroup comparison of the variations between the control and NFT groups assessing pre and post-scores confirmed these findings. On the other hand, intragroup score analysis showed that the minor variations found in control group were not significant for any of the cognitive functions. By contrast, intra-group analysis in NFT group showed changes in all cognitive functions except attention.

SNV was computed to provide a measure of global changes across cognitive functions. Results from analyzing this parameter were consistent with the differences showed in each single function. Since only attentional differences did not reach statistical significant significance, the global measure of variations reinforces the global tendency. Hence, according to these results, cognitive improvements in the visuospatial, language, memory and intellectual functions could be linked with NFT.

Two of the most widely used methods to stimulate certain frequencies or brain regions in NFT-related studies are evoking emotions or presenting emotional faces (Güntekin and Basar 2007). Although motor imagery strategies were also used for neurofeedback approaches in different diseases as autism (Pineda et al. 2014), MI-BCI is relatively novel to evaluate cognitive changes with elderly users. By means of the use of a BCI, a friendly environment is presented to the real users. Hence, participants were more predisposed to perform the training. Furthermore, it appeared to be an efficient strategy, since scores in several cognitive functions increased after only five sessions. A similar study that did not use MI-BCI needed more than 30 sessions to achieve statistically significant increments in the measured cognitive functions (Wang and Hsieh 2013).

In summary, cognitive variations in elderly people were obtained through repetitive stimulation of brain activity. Our novel NFT protocol based on MI-BCI yielded significant increases in real users across several cognitive features. All assessed cognitive functions showed enhanced Luria-AND scores, and most of these changes were significant. These results were obtained from a large group of users compared with typical BCI studies. Our results suggest that it is possible to improve cognitive performance in elderly people by means of NFT in few sessions. This study reinforces previous findings, while opening up the possibility of designing new NFT protocols based on BCIs.

Outlook

NFT using MI-BCI focused on elderly people is still an open issue. We have shown that it is possible to improve performance of different cognitive functions in elderly people with only five sessions of NFT using BCI technology. Nevertheless, it would be desirable to extend the population under study before obtaining generalizable and robust conclusions. In addition, further analyses showing potential changes in the EEG are needed. Future studies should also explore possible long-term effects of such training. This research could determine whether any cognitive improvements are transient or sustained over time. This is of paramount importance to the population that is focused this study: elderly people.

Overall, our results support our view that NFT performed with MI-BCI is a promising method for promoting brain plasticity and enhancing cognitive performance in elderly people. However, additional studies are necessary to further improve and explore our BCI and NFT protocol.

References

F.I.M Craik, T.A Salthouse, *Handbook of Aging and Cognition II*. (Psychology Press, United Kingdom, 2011)

A.L. Christensen, A practical application of the Luria methodology. J. Clin. Exp. Neuropsychol. **1**(3), 241–247 (1979)

United Nations, Department of Economic and Social Affairs, Population Division (2013). World Population Ageing 2013. ST/ESA/SER.A/348, http://www.un.org/en/development/desa/population/publications/pdf/ageing/WorldPopulationAgeing2013.pdf. Accessed 20 Nov 2014

J.J. Daly, J.R. Wolpaw, Brain-computer interfaces in neurological rehabilitation. Lancet Neurol. **7**(11), 1032–1043 (2008)

M. Arns, H. Heinrich, U. Strehl, Evaluation of neurofeedback in ADHD: the long and winding road. Biol. Psychol. **95**, 108–115 (2014)

M.B. Sterman, T. Egner, Foundation and practice of neurofeedback for the treatment of epilepsy. Appl. Psychophysiol. Biofeedback **31**, 21–35 (2006)

R. Coben, M. Linden, T.E. Myers, Neurofeedback for autistic spectrum disorder: a review of the literature. Appl. Psychophysiol. Biofeedback **35**, 83–105 (2010)

K.E. Thornton, D.P. Carmody, Efficacy of traumatic brain injury rehabilitation: Interventions of qEEG-guided biofeedback, computers, strategies, and medications. Appl. Psychophysiol. Biofeedback **33**(2), 101–124 (2008)

D. Vernon, T. Egner, N. Cooper, T. Compton, C. Neilands, A. Sheri, J. Gruzelier, The effect of training distinct neurofeedback protocols on aspects of cognitive performance. Int. J. Psychophysiol. **47**, 75–85 (2003)

S.M. Staufenbiel, A.M. Brouwer, A.W. Keizer, N.C. VanWouwe, Effect of beta and gamma Neurofeedback on memory and intelligence in the elderly. Biol. Psychol. **95**, 74–85 (2014)

J.R. Wang, S. Hsieh, Neurofeedback training improves attention and working memory performance. Clin. Neurophysiol. **124**(12), 2406–2420 (2013)

E. Angelakis, S. Stathopoulou, J.L. Frymiare, D.L. Green, J.L. Lubar, J. Kounios, EEG neurofeedback: a brief overview and an example of peak alpha frequency training for cognitive enhancement in the elderly. Clin. Neuropsychol. **21**, 110–129 (2006)

T. Ros, M.A. Munneke, D. Ruge, J.H. Gruzelier, J.C. Rothwell, Endogenous control of waking brain rhythms induces neuroplasticity in humans. Eur. J. Neurosci. **31**(4), 770–778 (2010)

J. Ghaziri, A. Tucholka, V. Larue, M. Blanchette-Sylvestre, G. Reyburn, G. Gilbert, J. Lévesque, M. Beauregard, Neurofeedback training induces changes in white and gray matter. Clin. EEG Neurosci. **44**(4), 265–272 (2013)

H.H. Jasper, Report of committee on methods of clinical examination in electroencephalography. Electroenceph. Clin. Neurophysiol. **10**, 370–375 (1958)

G. Schalk, D.J. McFarland, T. Hinterberger, N. Birbaumer, J.R. Wolpaw, BCI2000: a general-purpose brain-computer interface (BCI) system. IEEE Trans. Biomed. Eng. **51**, 1034–1043 (2004)

J.A. Pineda, D.S. Silverman, A. Vankov, J. Hestenes, Learning to control brain rhythms: making a brain-computer interface possible. IEEE Trans. Neural Syst. Rehabil. Eng. **11**(2), 181–184 (2003)

J.A. Pineda, E.V. Friedrich, K. LaMarca, Neurorehabilitation of social dysfunctions: a model-based neurofeedback approach for low and high-functioning autism. Front. Neuroeng. **7**, 29 (2014)

B. Güntekin, E. Basar, Emotional face expressions are differentiated with brain oscillations. Int. J. Psychophysiol. **64**, 91–100 (2007)

Airborne Ultrasonic Tactile Display BCI

**Katsuhiko Hamada, Hiromu Mori, Hiroyuki Shinoda
and Tomasz M. Rutkowski**

Abstract This chapter presents results of our project, which studied whether contactless and airborne ultrasonic tactile display (AUTD) stimuli delivered to a user's palms could serve as a platform for a brain computer interface (BCI) paradigm. We used six palm positions to evoke combined somatosensory brain responses to implement a novel contactless tactile BCI. This achievement was awarded the top prize in the Annual BCI Research Award 2014 competition. This chapter also presents a comparison with a classical attached vibrotactile transducer-based BCI paradigm. Experiment results from subjects performing online experiments validate the novel BCI paradigm.

Introduction

State-of-the-art brain computer interfaces (BCIs) are usually based on mental, visual or auditory paradigms, as well as body movement imagery paradigms, which require extensive user training and good eyesight or hearing. In recent years, alternative solutions have been proposed to make use of the tactile modality (Muller-Putz et al. 2006; Brouwer and Van Erp 2010; Mori et al. 2012) to enhance

---

K. Hamada is currently with DENSO Corporation, Japan.

K. Hamada · H. Shinoda
The University of Tokyo, Tokyo, Japan

H. Mori · T.M. Rutkowski (✉)
Life Science Center of TARA, University of Tsukuba, Tsukuba, Japan
e-mail: tomek@bci-lab.info
URL: http://bci-lab.info/

T.M. Rutkowski
RIKEN Brain Science Institute, Wako-shi, Japan

© The Author(s) 2015
C. Guger et al. (eds.), *Brain-Computer Interface Research*,
SpringerBriefs in Electrical and Computer Engineering,
DOI 10.1007/978-3-319-25190-5_6

BCI efficiency. The concept reported in this chapter further extends the brain's somatosensory channel by applying a contactless stimulus generated with an airborne ultrasonic tactile display (AUTD) (Iwamoto et al. 2008). This is an expanded version of a conference paper published by the authors (Hamada et al. 2014).

The rationale behind the use of the AUTD is that, due to its contactless nature, it allows for a more hygienic application, avoiding the occurrence of skin ulcers (bedsores) in patients in a locked-in state (LIS). The AUTD permits a less complex application of the BCI for the caregivers comparing to classical attached vibrotactile transducers' setups.

This chapter reports very encouraging results with AUTD-based BCI (autdBCI) compared to the classical paradigm using vibrotactile transducer-based oddball (P300 response-based) somatosensory stimulus (vtBCI) attached to the user's palms (Mori et al. 2012).

The rest of the chapter is organized as follows. The next section introduces the methods used in the study. The results obtained in online experiments with 13 healthy BCI users are then discussed. Finally, conclusions are drawn and directions for future research are outlined.

Methods

Thirteen male volunteer BCI users participated in the reported in this chapter experiments. The users' mean age was 28.54, with a standard deviation of 7.96 years. The experiments were performed at the Life Science Center of TARA, University of Tsukuba, at the University of Tokyo and at RIKEN Brain Science Institute, Japan. The online (real-time) EEG autdBCI and vtBCI paradigm experiments were conducted in accordance with the *WMA Declaration of Helsinki-Ethical Principles for Medical Research Involving Human Subjects* and the procedures were approved and designed in agreement with the ethical committee guidelines of the Faculty of Engineering, Information and Systems at University of Tsukuba, Japan (experimental permission 2013R7).

The AUTD stimulus generator produced vibrotactile contactless stimulation of the human skin via the air using focused ultrasound (Iwamoto et al. 2008; Hamada 2014). The effect was achieved by generating an ultrasonic radiation static force produced by intense sound pressure amplitude (a nonlinear acoustic phenomenon). The radiation pressure deformed the surface of the skin on the palms, creating a virtual touch sensation. An array of ultrasonic transducers mounted on the AUTD (see Fig. 1) created the focused radiation pressure at an arbitrary focal point by choosing a phase shift of each transducer appropriately (the so-called phased array technique). Modulated radiation pressure created a sensation of tactile vibration similar to the one delivered by classical vibrotactile transducers attached to the user's palms, as shown in Fig. 2. The AUTD device developed by the authors (Iwamoto et al. 2008; Hamada et al. 2014) (see Fig. 1) adhered to ultrasonic medical standards and did not exceed the permitted skin absorption levels (approximately 40 times

Fig. 1 The AUTD array with ultrasonic transducers used to create the contactless tactile pressure sensation

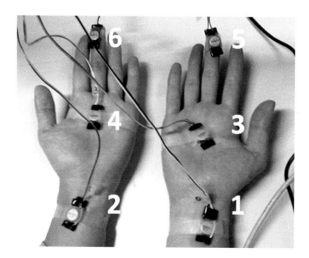

Fig. 2 User's palms with attached vibrotactile transducers used in vtBCI experiments. Each stimulus location reflects a different digit

below the permitted limits). The effective vibrotactile sensation was set to 50 Hz (Hamada 2014) to match with tactile skin mechanoreceptors' frequency characteristics and the notch filter that EEG amplifiers use for power line interference rejection.

As a reference, in the second vtBCI experiment, contact vibrotactile stimuli were also applied to locations on the users' palms via the transducers HIHX09C005-8. Each transducer in the experiments was set to emit a square acoustic frequency wave at 50 Hz, which was delivered from the ARDUINO micro-controller board with a custom battery-driven and isolated power amplifier and software developed in-house and managed from a *MAX 6* visual programming environment.

The two experiment set-ups above are presented in Figs. 2 and 3. Two types of experiments were performed with the volunteer healthy users. Experiments with the target paralyzed users are planned as a followup of the current pilot project. Psychophysical experiments with foot-button-press responses were conducted to test uniform stimulus difficulty levels from response accuracy and time measurements. The subsequent tactile oddball online BCI EEG experiments evaluated the autdBCI paradigm efficiency and allowed for a comparison with the classical skin contact-based vtBCI reference. In both the above experiment protocols, the users were instructed to spell sequences of six digits representing the stimulated positions on their palms. The training instructions were presented visually by means of the *BCI2000* (Schalk et al. 2000) and *MAX 6* programs with the numbers 1–6 representing the palm locations as depicted in Fig. 2.

The EEG signals were recorded with the g.USBamp amplifier system from g.tec Medical Engineering GmbH, Austria, using 16 active g.LADYbird electrodes. The electrodes were attached to the head locations: *Cz, Pz, P3, P4, C3, C4, CP5, CP6,*

Fig. 3 A user during the autdBCI experiment with both palms placed under the AUTD array with ultrasonic transducers

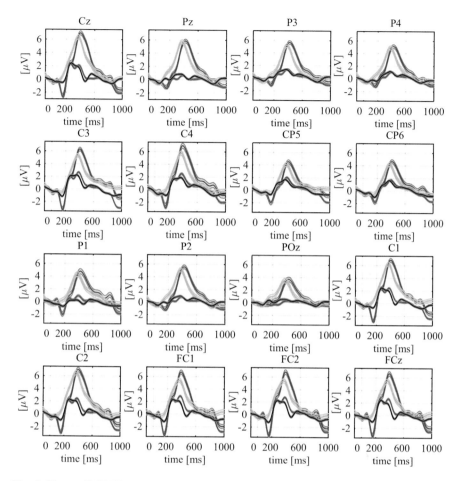

Fig. 4 The autdBCI (*blue* targets; *red* non-targets) and vtBCI (*green* targets; *black* non-targets) grand mean averaged ERP responses, together with standard *error bars*

P1, P2, POz, C1, C2, FC1, FC2, and *FCz,* as in the 10/10 extended international system. The ground electrode was attached to the *FPz* position, and the reference was attached to the left earlobe. No electromagnetic interference was observed from the AUTD or vibrotactile transducers operating with frequencies notch-filtered together with power line interference from the EEG. The EEG signals captured were processed online with an in-house extended BCI2000-based application (Schalk et al. 2000), using a stepwise linear discriminant analysis (SWLDA) classifier (Krusienski et al. 2006) with features drawn from 0 to 800 ms ERP intervals decimated by a factor of 20.

The stimulus length and inter-stimulus-interval were set to 400 ms, and the number of epochs to average was set to 15. The EEG recording sampling rate was set at 512 Hz, and the high and low pass filters were set at 0.1 and 60 Hz,

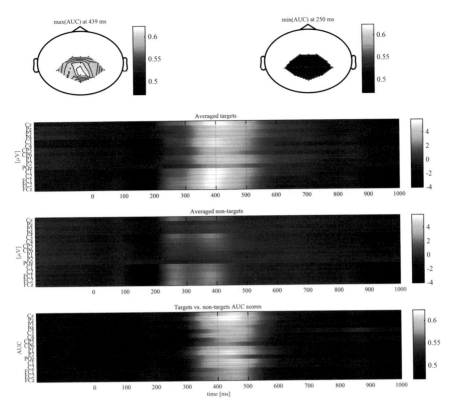

Fig. 5 The autdBCI grand mean averaged ERP responses, shown as matrix plots for targets in the *top panel*; non-targets in the middle; and area under the curve (AUC) of the response discriminability analysis (AUC > 0.5 marks the discriminable latencies)

respectively. The notch filter to remove power line interference removed activity between 48–52 Hz. Each user performed three experiment runs (randomized 90 targets and 450 non-targets each). As feedback, the spelled numbers (palm position assigned digits as in Fig. 2) were shown on a display to the user.

Results

The grand mean averaged evoked responses to targets and non-targets are depicted together with standard error bars in Fig. 4 and as matrices with an area under the curve (AUC) analysis for feature separability in Fig. 5. The BCI six digit sequences spelling accuracy analyses for both experiments for the various averaging options are summarized in Fig. 6. The chance level was 16.6 %. The mean six digit sequence spelling accuracies for 15-trial averaged ERPs were 63.8 and 69.4 % for

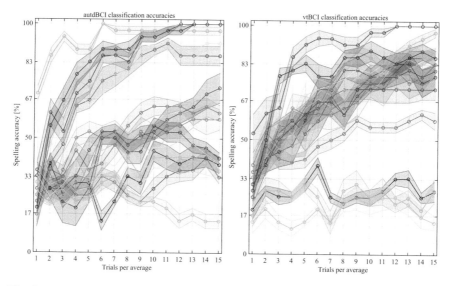

Fig. 6 Averaged autdBCI and vtBCI spelling accuracy across six digits, colour coded for each user with standard *error bars*

autdBCI and vtBCI, respectively. The maximum accuracies were 78.3 and 84.6 % respectively. The differences were not significant, supporting the concept of using autdBCIs. However, a single trial classification offline analysis of the collected responses resulted with the best obtained accuracies of 83.0 % for autdBCI and 53.8 % for vtBCI, leading to a possible 19.2 bit/min and 7.9 bit/min, respectively.

In the case of the autdBCI, only a single user's results were bordering on the level of chance, and four subjects attained 100 % (10 trials averaging). On average, lower accuracies were obtained with the classical vtBCI, with which three users bordered on the level of chance, and only one user scored 100 % accuracy level in SWLDA-classified averaged responses.

Conclusions

This study demonstrates results obtained with a novel six-command-based autdBCI paradigm. We compared the results with a classical vibrotactile transducer stimulus-based paradigm. The experiment results obtained in this study confirm the validity of the contactless autdBCI for interactive applications and the possibility to further improve the results through single trial-based SWLDA classification.

The EEG experiment with our paradigm confirms that contactless (airborne) tactile stimuli can be used to create six command-based BCIs in real time. A short demo with online application of the paradigm to robotic arm control is available on YouTube (Rutkowski et al. 2014).

The results presented offer a step forward in developing and validating novel neurotechnology applications. Since most users did not achieve very high accuracy during online BCI operation, especially with only a few trials, the current paradigm obviously requires improvement and modification. These requirements determine the major lines of study for future research.

However, even in its current form, the proposed autdBCI can be regarded as a practical solution for LIS patients (locked into their own bodies despite often intact cognitive functioning), who cannot use vision or auditory-based interfaces due to sensory or other disabilities. The reported autdBCI project was awarded The BCI Annual Research Award 2014 for "A fascinating new idea never explored before," according to the Chairman of the Jury for the Annual BCI Research Award 2014, Prof. Gernot R. Mueller-Putz from the Institute for Knowledge Discovery, Graz University of Technology, Austria.

Acknowledgments H. Mori and T.M. Rutkowski were supported in part by the Strategic Information and Communications R&D Promotion Program No. 121803027 of The Ministry of Internal Affairs and Communication in Japan.

References

A.-M. Brouwer, J.B.F. Van Erp, A tactile P300 brain-computer interface. Frontiers Neurosci. **4**(19) (2010) [Online]. http://www.frontiersin.org/neuroprosthetics/10.3389/fnins.2010.00019/abstract

K. Hamada, Brain-computer interface using airborne ultrasound tactile display. Master Thesis, The University of Tokyo, Tokyo, Japan (2014) (in Japanese)

K. Hamada, H. Mori, H. Shinoda, T.M. Rutkowski, Airborne ultrasonic tactile display brain-computer interface paradigm. In *Proceedings of the 6th International Brain-Computer Interface Conference 2014*, eds. by G. Mueller-Putz, G. Bauernfeind, C. Brunner, D. Steyrl, S. Wriessnegger, R. Scherer. 1em plus 0.5em minus 0.4em Graz University of Technology Publishing House (2014). Article ID 018-1-4. [Online]. http://castor.tugraz.at/doku/BCIMeeting2014/bci2014_018

T. Iwamoto, M. Tatezono, H. Shinoda, Non-contact method for producing tactile sensation using airborne ultrasound. In *Haptics: Perception, Devices and Scenarios*, ser. Lecture Notes in Computer Science, ed. by M. Ferre 1em plus 0.5em minus 0.4em vol 5024 (Springer, Berlin, 2008) pp. 504–513. [Online]. http://dx.doi.org/10.1007/978-3-540-69057-3_64

D.J. Krusienski, E.W. Sellers, F. Cabestaing, S. Bayoudh, D.J. McFarland, T.M. Vaughan, J.R. Wolpaw, A comparison of classification techniques for the P300 speller. J. Neural Eng. **3**(4), 299 (2006). [Online]. http://stacks.iop.org/1741-2552/3/i=4/a=007

H. Mori, Y. Matsumoto, S. Makino, V. Kryssanov, T.M. Rutkowski, Vibrotactile stimulus frequency optimization for the haptic BCI prototype. In *Proceedings of The 6th International Conference on Soft Computing and Intelligent Systems, and the 13th International Symposium on Advanced Intelligent Systems*, Kobe, Japan, November 20–24 (2012) pp. 2150–2153. [Online]. http://arxiv.org/abs/1210.2942

G. Muller-Putz, R. Scherer, C. Neuper, G. Pfurtscheller, Steady-state somatosensory evoked potentials: suitable brain signals for brain-computer interfaces? IEEE Trans. Neural Syst. Rehab. Eng. **14**(1), 30–37 (2006)

T.M. Rutkowski, The autdBCI and a robot control (the winner project of The BCI Annual Research Award 2014). YouTube video. [Online]. http://youtu.be/JE29CMluBh0

G. Schalk, J. Mellinger, A practical guide to brain-computer interfacing with BCI2000. 1em plus 0.5em minus 0.4em (Springer, London, 2010)

Heterogeneous BCI-Triggered Functional Electrical Stimulation Intervention for the Upper-Limb Rehabiliation of Stroke Patients

Jaime Ibáñez, J.I. Serrano, M.D. Del Castillo, E. Monge, F. Molina and J.L. Pons

Abstract Stroke motor rehabilitation strategies using neuromodulation paradigms that take advantage of the motor predictive characteristics of the electroencephalographic signal are currently subject to extensive research. Such rehabilitation strategies follow a top-down approach, in which targeted neurophysiological changes in the central nervous system are expected to induce functional improvement. This chapter presents a series of studies regarding processing algorithms to detect motor intentionality and a neuromodulation paradigm to improve the upper-limb functionality. The experiments were developed and tested with stroke patients.

Keywords Movement-related cortical potentials · Event-related desynchronization · Neuromodulation · Brain-computer interfaces · Stroke neurorehabilitation

Introduction

Stroke is defined as a sudden, focal neurological deficit due to a cerebrovascular abnormality. It is a leading cause of long-term motor disability among adults, with increasing prevalence during the last decades. Approximately one third of all stroke survivors become chronic patients with a permanent disability affecting at least one

J. Ibáñez (✉) · J.L. Pons
Neural Rehabilitation Group, Cajal Institute, Spanish National Research Council (CSIC), 28002 Madrid, Spain
e-mail: jaime.ibanez@csic.es

J.I. Serrano · M.D. Del Castillo
Neural and Cognitive Engineering Group, Centro de Automática y Robótica, CSIC, 28500 Arganda del Rey, Spain

E. Monge · F. Molina
LAMBECOM Group, Health Sciences Faculty, Universidad Rey Juan Carlos, Alcorcón, Spain

© The Author(s) 2015
C. Guger et al. (eds.), *Brain-Computer Interface Research*,
SpringerBriefs in Electrical and Computer Engineering,
DOI 10.1007/978-3-319-25190-5_7

part of the body, despite intensive rehabilitation. Traditional approaches in the rehabilitation of stroke patients are unsuccessful in a significant number of cases, especially in relation to upper-limb function recovery (Lai et al. 2002). After the initial 6–12 months post-stroke, patients are considered to be in a chronic stage and, from that moment on, significant motor recovery is rare. Given age-based risk of stroke and the rapidly aging population of developed countries, successful rehabilitation strategies for stroke patients with diminished functional capacities will have huge public health implications in the short and long terms (Belda-Lois et al. 2011; Dobkin 2004).

During the last decades, different groups have experimentally demonstrated either sensitization or conditioning paradigms producing long-term synaptic changes at supraspinal (Brus-ramer et al. 2007; Stefan et al. 2000) and spinal (Thompson et al. 2013) levels that may in turn lead to motor readaptations. Currently available noninvasive techniques allowing the electrophysiological acquisition and stimulation of specific structures in the central nervous system— either through central or peripheral pathways—make it possible to develop such patient-specific neuromodulation strategies transferable to the clinical practice. These guided and targeted interventions have been proposed to trigger neurophysiological changes restoring specific motor functions in humans. In this regard, the electroencephalographic (EEG) signal becomes a valuable source of information, since it allows the real-time characterization of movement-related cortical processes directly linked with the planning and execution of voluntary movements (Pfurtscheller et al. 2003). Indeed, a number of EEG-based applications have been presented during the last decade in neuromodulation paradigms (Niazi et al. 2012; Ramos Murguialday et al. 2013). The characteristics of the EEG in stroke patients are thoroughly described in the literature (Serrien et al. 2004; Stepien et al. 2010) and the possibilities of decoding motor intentions in this population have been explored (Ramos Murguialday et al. 2013). It is therefore expected that the role of EEG-based technologies in neuromodulation paradigms to improve the motor function of stroke survivors will receive significant interest.

We have been developing and validating EEG-based techniques to accurately detect upper-limb motor intentions in patients who have suffered a stroke. The ultimate goal is to validate a neuromodulation platform based on EEG and functional electrical stimulation (FES) technologies to selectively improve the upper-limb function of stroke patients. This chapter provides an experimental review of the application of EEG systems in rehabilitation interventions for patients with a stroke affecting their motor function. Additionally, a summary of some recently obtained experimental data regarding the validation of the proposed BCI intervention with the targeted population is included.

Neuromodulation in Stroke Patients—BCI Perspective

Rehabilitation is a therapeutic process that aims to maximize the physical, psychological and social potential of the patient (Muro et al. 2000). Although many people manage to partially recover the motor function of the lower extremity after a stroke, most patients do not use the upper extremity in activities of daily living after months of rehabilitation (Adams et al. 1994). Functional recovery is observed in the upper extremity in less than 15 % of the subjects (Hendricks et al. 2002).

Although the origin of neurologic disabilities is located centrally, conventional therapies have traditionally focused on providing sensory feedback (Murphy and Corbett 2009) and performing real movements in the affected limbs of the patients. In this sense, they have been based in a bottom-up approach, i.e. the rehabilitation focuses on the peripheral function, which is in turn expected to induce central neurophysiological changes. Nevertheless, the principal mechanisms implicated in the motor recovery of stroke survivors involve enhanced activity of the primary motor cortex induced by active motor training (Calautti and Baron 2003). While peripheral stimulation has not proven to be a locally specific way of promoting plastic changes, an induced coherent activation of sensory feedback circuits and structures in the primary motor cortex is expected to reinforce corticomuscular connections according to Hebbian learning principles, and thus support functional recovery (Murphy and Corbett 2009). Taking into account that the connection between the sensorimotor cortex and peripheral muscles in stroke survivors has been altered, this rehabilitation strategy appears to be a logical step to reinforce the corticomuscular descending pathways to regain motor control.

This implies switching to a top-down rehabilitation strategy in which the mechanisms targeted for modification through rehabilitation are the central structures of the nervous system in charge of the movement generation. In this concept, the peripheral rehabilitation is carried out in synchrony with the activity of the functionally associated structures of the brain, or rather triggered by it. The coupling promotes a cause-effect action from intention to execution of movement, thus increasing the associative facilitation of efferent pathways. Indeed, experimental paradigms using Paired Associative Stimulation (PAS, i.e. the application of timely associated electrical stimuli in cortical and muscular regions) have proven to be an effective way to strengthen the corticomuscular connections (Stefan et al. 2000).

Most BCI-based rehabilitation systems for stroke survivors that have been proposed so far have been based on this idea of using brain activity to drive external systems that provide an associative proprioceptive feedback and sometimes assistance. This sensory feedback may induce plasticity underlying the restoration of normal motor control. The basis of this approach is that activity-dependent CNS plasticity can induce changes at synaptic, neuronal and circuits levels in cortical and subcortical structures of the brain and so produce a more normal motor control (Nudo 2006). To date, only a few studies have been able to demonstrate actual neurophysiological supraspinal changes in subjects undergoing a BCI intervention for stroke rehabilitation (Niazi et al. 2012; Ramos Murguialday et al. 2013; Várkuti et al. 2013).

Among them, the studies presented by Mrachacz-Kersting and colleagues (Mrachacz-Kersting et al. 2012; Niazi et al. 2012; Xu et al. 2014) represent a specially relevant contribution in the field. Their studies highlighted the critical relevance of the temporal association between the cortical activity associated with the movement initiation and the peripheral activity, to produce actual functional and plastic changes that could be observed after a single experimental session consisting of around 50–60 movement trials. Most BCI studies based on EEG-SMR present a longer delay between motor intention and movement onset detection, which is a probable cause of insufficient single-session neurophysiological changes in patients because of these interventions. For this reason, the proposed EEG-based processing algorithms shown here stress the need of temporal accuracy in motor intention detections.

EEG-Based Detection of the Movement Intentions with Time Precision

According to the previously mentioned main goals of most BCI systems for stroke rehabilitation, the reliable decoding of movement-related cortical events with time precision becomes a key aspect. Cortical patterns measured with EEG may be classified according to the cause that generates them, either as a consequence of external stimuli or as a result of self-initiated mental processes. Cortical patterns of the second class can be in turn classified according to their morphology and the frequency components that conform them in two groups: slow cortical patterns, characterized by positive or negative deflections of the EEG low-frequency components (<1 Hz), and patterns associated with changes in the power of EEG cortical rhythms. The Event-Related Desynchronization (ERD) and the Readiness- or Bereitschaftspotential (BP) are the two best documented movement-related cortical patterns, both appearing as a consequence of motor planning and processing. On the one hand, the movement-related ERD is characterized by the decay in the sensorimotor rhythms of cortical areas associated with the part of the body experiencing or involved in a motor task (see Fig. 1a) (Pfurtscheller and da Silva 1999). The ERD decay can be already observed up to 2 s before a voluntary movement begins. On the other hand, the BP belongs to the slow cortical components of the EEG activity and it consists in a decrease of the EEG amplitude starting ~1.5 s before self-initiated movements with a steeper decay during the last 200–300 ms (see Fig. 1b). The supplementary area, the premotor cortex and the motor areas around the central sulcus are the regions where the BP shows larger amplitudes (Shibasaki and Hallett 2006). Both BP and ERD have been extensively used to detect online motor processing states (Townsend et al. 2004), to anticipate the moments when voluntary movements begin (Ibáñez et al. 2013), to decode movement parameters such as velocity or strength (Jochumsen et al. 2013), or to distinguish between different classes of movements to be performed (Pfurtscheller et al. 2006).

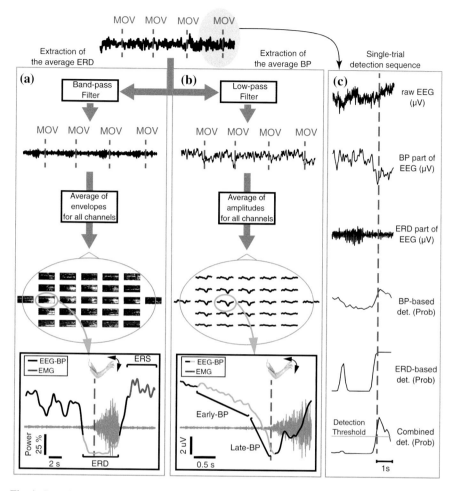

Fig. 1 Description of the extraction process for ERD (**a**) and BP (**b**) average cortical patterns from the raw EEG signal. Schematic representation (**c**) of the single-trial ERD + BP-based detection of the movement onset according to the classification process proposed in this chapter

A number of studies have proposed the online characterization of the BP pattern to detect movement intentions in humans with high temporal precision (Garipelli et al. 2013; Jochumsen et al. 2013; Lew et al. 2012; Niazi et al. 2011; Xu et al. 2014). Since the BP presents an identifiable pattern that decays until the movement starts, it is suitable to achieve temporal precision in the detection of the onsets of voluntary movements. In fact, previous studies showing results of online systems based on this pattern indicate that average detection latencies of 315 ± 165 ms can be obtained (Xu et al. 2014). Nevertheless, the BP is not detectable in all cases, since some subjects do not present a significant pattern during self-paced movements, especially subjects who do not have the capacity to perform movements at a normal speed (Jochumsen et al. 2013). In fact, altered BP patterns have been observed in previous studies with stroke patients (Daly et al. 2006; Fang et al. 2007).

A possible way of boosting EEG-based systems that detect voluntary movement onsets is to combine the BP with other EEG movement-related patterns providing complementary information, such as the ERD (Fatourechi et al. 2008). Although a variable anticipation may be given in the ERD for a specific channel/frequency combination (see lower panel in Fig. 1a), the spatio-tempo-frequential distribution of the ERD observed when averaging a number of EEG segments preceding voluntary movements shows a desynchronization pattern attached to the movement event (Bai et al. 2005). Therefore, the analysis of the ERD also provides relevant information regarding the timing of volitional motor actions. As in the analysis of the BP, the ERD pattern of stroke patients presents variations with respect to healthy subjects (Stepien et al. 2010). Therefore, it is of special relevance to study how stroke-related cortical changes may affect a BCI driven by these cortical patterns.

According to these previous lines regarding the suitability of the ERD and BP cortical patterns to characterize online and with temporal precision movement events, we developed a BCI system taking advantage of the complementarity of the information they provide (Ibáñez et al. 2014). Our approach classified the ERD and BP information separately and then combined the classification. A Naïve Bayes classifier with 10 features was used to detect the ERD pattern preceding the movements. Band-pass filtering (between 6 and 30 Hz) and a small Laplacian filter were applied in the pre-processing stage. Frontal, fronto-central, central, centro-parietal and parietal channels were considered and the features fed to the classifier were power estimations between 7–30 Hz. The 10 features (channel-frequency pairs) used by the classifier were selected using the Bhattacharyya distance. To detect the BP, we first applied a finite impulse response band-pass filter with linear phase (FIR filter, 15th order, $0.05 \text{ Hz} < \text{f1}$, $1 \text{ Hz} > \text{f2}$) to extract the low-frequency component of the EEG signal. Three virtual channels were obtained by subtracting the average potential of channels F3, Fz, F4, C3, C4, P3, Pz and P4 to channels C1, Cz and C2. The average BP was computed for the three channels using the training data, and the optimal one (with the highest absolute BP peak) was selected for the BP-based detection of movement onsets. A matched filter was then designed with the training data by computing the average BP in the time interval $\{-1.5; 0 \text{ s}\}$. This matched filter was then applied to the online data, generating maximal outputs when the low-frequency EEG signal in the analysed virtual channel resembled that of the BP pattern. Finally, outputs from ERD-based and BP-based detectors were combined using a logistic regression classifier. Training examples of the resting condition were taken from estimations of the two detectors between -3 and -0.5 s with respect to the movement onset. The output estimations of the ERD and the BP classifiers at the movement onset were used to model the movement state. A schematic example of a single trial detection of a movement onset based on this whole procedure is shown in Fig. 1c.

A New BCI-FES Intervention for the Upper-Limb of Chronic Stroke Patients

In recent studies, we tested the ability of the previously proposed detector of motor intentions in chronic stroke patients while they performed self-initiated reaching movements with the affected arm. There were two main goals: (1) to validate the potential benefits of combining ERD- and BP-based systems to locate upper-limb voluntary motor events in chronic stroke patients, and (2) to analyse the effects of a neuromodulation intervention with the proposed EEG-based detector and FES.

To validate the EEG-based detection of motor intentions combining ERD and BP information, a first offline experiment was carried out in which patients were asked to perform self-paced movements without afferent feedback. On average, 82.2 ± 10.4 % of the movements were correctly detected (true positives or TP) by the system combining ERD and BP information, and 1.52 ± 1.89 false detections (false positives or FP) were generated per minute. The detections were performed with latencies (Lat.) of 35.9 ± 352.3 ms with respect to the actual movements (located with gyroscopes). Interestingly, the detector accuracy was similar to that observed with healthy subjects (TP = 74.5 ±10.8 %, FP/min = 1.32 ± 0.87), although detection latencies were longer (Lat. = 89.9 ± 349.2 ms). These differences in the detection latencies were associated with differences in the average temporal patterns of the ERD and BP between patients and healthy subjects: the peak of the BP (see Fig. 2) and the onset of the ERD were typically delayed in patients compared to controls. This is also in line with previous observations of altered movement-related cortical patterns in patients who have suffered a stroke (Daly et al. 2006; Serrien et al. 2004; Stepien et al. 2010). When comparing the performance of the combined (BP + ERD) detector with detectors either relying on the ERD or BP patterns alone, this combination of sources of information significantly improved detection accuracy in the patients. The percentage of good trials (GT, the percentage of movement examples in which no FPs were generated during the preceding resting period and a correct movement detection was achieved) was

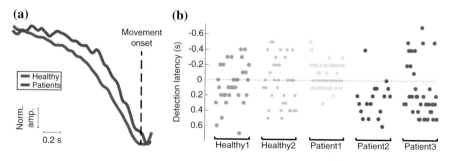

Fig. 2 a Average BP (among 6 healthy subjects and 6 patients) showing a delay in the patients' BP maximum peak. **b** Locations of the EEG-based movement onset detections with respect to the actual movement onsets (t = 0 s), from 2 healthy subjects and 3 patients

increased in 13.3 ± 10.9 and 12.6 ± 16.3 % when comparing the combined detector with the BP- and ERD-based detectors, respectively.

The next step in the validation of the system in a conditioning paradigm was to test its performance online while patients received electrical afferent stimuli below the motor threshold when motor intentions were detected. This was performed with two healthy subjects and three chronic stroke patients who had participated in the previous study. In this case, instead of asking the participants to perform movements at their own chosen pace, patients were instructed to wait until an enabling signal was given to them, at which time they were supposed to wait a few seconds (3–5 s) and perform a self-initiated movement. The enabling signal was programmed to appear after each movement had been performed and as soon as the cortical activity showed a resting condition (according to the ERD and BP patterns). This paradigm ensured that a true basal condition was given at the beginning of each trial and required the patients to concentrate in reaching a resting state condition before starting each new movement, which in turn was expected to empower the system's performance by reducing the amount of FP. The detection results in the patients showed lower TP (66.9 ± 26.4 %) and especially FP/min (0.42 ± 0.17) compared to the offline results presented before. This was likely caused by both an alteration in the level of the applied threshold in both cases and the effects of the used paradigm. The detection latencies were again close to the actual onsets of the movements (see Fig. 2b), although in one of the three patients, the detections were generally obtained after the actual movements in most cases (patient 2).

Finally, a feasibility study was carried out to analyse the possible beneficial effects in chronic stroke patients of a BCI intervention using the previously described EEG-based detector and FES. Four patients completed eight intervention sessions during 1 month. In each session, patients performed self-paced reaching movements with the affected upper-limb and FES was delivered every time a movement intention was detected by the EEG-based system. Overall, patients could reliably control the interface by naturally performing the movements and low detection latencies were obtained in most cases. For each intervention session and patient, the percentages of good trials obtained and the latencies of the detections were measured. The GT results across patients and sessions was 66.3 ± 15.7 %. The GT increased across sessions in three patients, suggesting learning mechanisms in the interaction with the BCI platform. Results in terms of TP and FP/min were comparable to those obtained in the first offline validation experiment. The detection latencies were stable across sessions and in two cases they slightly improved as the intervention evolved. The average detection latency (considering all sessions and patients) was 112 ± 278 ms (with the reference movement onset being located with data from gyroscopes). Unsuccessful results of the BCI system were only observed in one session with a patient. In that case, the BCI-based intervention was cancelled since the patient reported an uncomfortable interaction with the FES system. In all sessions, patients were also asked to imagine the movements instead of performing them to analyse the system's performance with these a priori weaker patterns. In this case, detection results slightly dropped, although they were comparable with the ones obtained with actual movements. Functional changes were assessed by comparing the Fügl-Meyer Index (FMI) measured before and after the experimental month. An average increase of 11.5 ± 5.5 points was observed. The measured FMI changes in three patients were above the minimal detectable change (which is 5.2 points for upper-extremity assessments). Changes of FMI in the fourth patient were slightly below this threshold, despite the positive results observed in the self-report test. Interestingly, this patient was also the one showing the worst BCI results in terms of detection latencies across all intervention sessions.

Current State and Future Perspectives

So far, BCI-based neuromodulation strategies for stroke rehabilitation have been able to address a number of relevant questions in the field. BCI technology that can be controlled by the cortical waves of patients with cortical lesions has been demonstrated (Buch et al. 2008; Ramos Murguialday et al. 2013). In addition, recent studies have advanced the methods to achieve reliable estimations of motor-related cortical states with temporal precision, which further boosts these neuromodulation applications (Ibáñez et al. 2014; Jochumsen et al. 2014). The neurophysiological and functional potential impacts of neuromodulation paradigms pairing the cortical activity at specific time instants with afferent stimuli, induced either electrically or physically/mechanically, has also been validated (Niazi et al. 2012; Xu et al. 2014). Taken together, these advances represent a strong background for subsequent studies in the near future, in which larger stroke populations will need to be recruited in clinical validation studies. This effort will further elucidate the interplay between the BCI reached performance in each person, the attainable neurophysiological changes induced (especially those associated with corticomuscular facilitation) and the functional improvement of the patients as a result of different rehabilitation intensities with these technologies. To achieve these goals, further developments in EEG acquisition systems and processing algorithms will need to be carried out so that the technology can be easily transferred to clinical practice. Additionally, further improvement of placebo-controlled conditions must be achieved to fully quantify the relevance of BCI technology in stroke rehabilitation.

References

H. Adams, T. Brott, R. Crowell, A. Burlan, C. Gomez, J. Grotta, E. Al, Guidelines for management of patient with acute ischemic stroke: a statement for healthcare profession from a special writing group of the stroke council. Stroke **25**, 1901–1914 (1994)

O. Bai, Z. Mari, S. Vorbach, M. Hallett, Asymmetric spatiotemporal patterns of event-related desynchronization preceding voluntary sequential finger movements: a high-resolution EEG study. Clin. Neurophysiol. **116**, 1213–1221 (2005)

J.M. Belda-Lois, S. Mena-del Horno, I. Bermejo-Bosch, J.C. Moreno, J.L. Pons, D. Farina, M. Iosa, F. Tamburella, A. Ramos Murguialday, A. Caria, T. Solis-Escalante, C. Brunner, M. Rea, Rehabilitation of gait after stroke: a review towards a top-down approach. J. Neuroeng. Rehabil. **8**, 66 (2011)

M. Brus-ramer, J.B. Carmel, S. Chakrabarty, J.H. Martin, Electrical stimulation of spared corticospinal axons augments connections with ipsilateral spinal motor circuits after injury **27**, 13793–13801 (2007)

E. Buch, C. Weber, L.G. Cohen, C. Braun, M.A. Dimyan, T. Ard, J. Mellinger, A. Caria, S. Soekadar, A. Fourkas, N. Birbaumer, Think to move: a neuromagnetic brain-computer interface (BCI) system for chronic stroke. Stroke **39**, 910–917 (2008)

C. Calautti, J.-C. Baron, Functional neuroimaging studies of motor recovery after stroke in adults: a review. Stroke **34**, 1553–1566 (2003)

J.J. Daly, Y. Fang, E.M. Perepezko, V. Siemionow, G.H. Yue, Prolonged cognitive planning time, elevated cognitive effort, and relationship to coordination and motor control following stroke. IEEE Trans. Neural Syst. Rehabil. Eng. **14**, 168–171 (2006)

B.H. Dobkin, Strategies for stroke rehabilitation. Lancet Neurol. **3**, 528–536 (2004)

Y. Fang, G.H. Yue, K. Hrovat, V. Sahgal, J.J. Daly, Abnormal cognitive planning and movement smoothness control for a complex shoulder/elbow motor task in stroke survivors. J. Neurol. Sci. **256**, 21–29 (2007)

M. Fatourechi, R.K. Ward, G.E. Birch, A self-paced brain-computer interface system with a low false positive rate. J. Neural Eng. **5**, 9–23 (2008)

G. Garipelli, R. Chavarriaga, R.J. del Millán, Single trial analysis of slow cortical potentials: a study on anticipation related potentials. J. Neural Eng. **10**, 036014 (2013)

H. Hendricks, M. Zwarts, E. Plat, J. van Limbeek, No TitleSystematic review for the early prediction of motor and functional outcome after stroke by using motor-evoked potentials. Arch. Phys. Med. Rehabil. **83**, 1303–1308 (2002)

J. Ibáñez, J.I. Serrano, M.D. del Castillo, J.A. Gallego, E. Rocon, Online detector of movement intention based on EEG. Application in tremor patients. Biomed. Signal Process. Control **8**, 822–829 (2013)

J. Ibáñez, J.I. Serrano, M.D. del Castillo, E. Monge-Pereira, F. Molina-Rueda, I. Alguacil-Diego, J. L. Pons, Detection of the onset of upper-limb movements based on the combined analysis of changes in the sensorimotor rhythms and slow cortical potentials. J. Neural Eng. **11**, 056009 (2014)

M. Jochumsen, I. Niazi, N. Mrachacz-Kersting, D. Farina, K. Dremstrup, Detection and classification of movement-related cortical potentials associated with task force and speed. J. Neural Eng. **10**, 056015 (2013)

M. Jochumsen, I. Niazi, H. Rovsing, C. Rovsing, Detection of movement intentions through a single channel of electroencephalography **7** (2014)

S.-M. Lai, S. Studenski, P.W. Duncan, S. Perera, Persisting consequences of stroke measured by the stroke impact scale. Stroke **33**, 1840–1844 (2002)

E. Lew, R. Chavarriaga, S. Silvoni, J.R. Millán, Detection of self-paced reaching movement intention from EEG signals. Front Neuroeng. **5**, 13 (2012)

N. Mrachacz-Kersting, S.R. Kristensen, I. Niazi, D. Farina, Precise temporal association between cortical potentials evoked by motor imagination and afference induces cortical plasticity. J. Physiol. **590**, 1669–1682 (2012)

J. Muro, J. Pedro-Cuesta, J. Almazan, W. Holmqvist, Stroke recovery in South Madrid. Function and motor recovery, resource utilization, and family support. Stroke **31**, 1352–1359 (2000)

T.H. Murphy, D. Corbett, Plasticity during stroke recovery: from synapse to behaviour. Nat. Rev. Neurosci. **10**, 861–872 (2009)

I. Niazi, N. Jiang, O. Tiberghien, J.F. Nielsen, K. Dremstrup, D. Farina, Detection of movement intention from single-trial movement-related cortical potentials. J. Neural Eng. **8**, 066009 (2011)

I. Niazi, N. Mrachacz-Kersting, N. Jiang, K. Dremstrup, D. Farina, Peripheral electrical stimulation triggered by self-paced detection of motor intention enhances motor evoked potentials. IEEE Trans. Neural Syst. Rehabil. Eng. **20**, 595–604 (2012)

R.J. Nudo, Mechanisms for recovery of motor function following cortical damage. Curr. Opin. Neurobiol. **16**, 638–644 (2006)

G. Pfurtscheller, F.H.L. da Silva, Event-related EEG/EMG synchronization and desynchronization: basic principles. Clin. Neurophysiol. **110**, 1842–1857 (1999)

G. Pfurtscheller, B. Graimann, J.E. Huggins, S.P. Levine, L.A. Schuh, Spatiotemporal patterns of beta desynchronization and gamma synchronization in corticographic data during self-paced movement. Clin. Neurophysiol. **114**, 1226–1236 (2003)

G. Pfurtscheller, C. Brunner, A. Schlögl, F.H.L. da Silva, F.H. Lopes da Silva, Mu rhythm (de) synchronization and EEG single-trial classification of different motor imagery tasks. Neuroimage **31**, 153–159 (2006)

A. Ramos Murguialday, D. Broetz, M. Rea, L. Läer, O. Yilmaz, F.L. Brasil, G. Liberati, M.R. Curado, E. Garcia-Cossio, A. Vyziotis, W. Cho, M. Agostini, E. Soares, S. Soekadar, A. Caria, L.G. Cohen, N. Birbaumer, Brain-machine interface in chronic stroke rehabilitation: a controlled study. Ann. Neurol. (2013)

D.J. Serrien, L.H.A. Strens, M.J. Cassidy, A.J. Thompson, P. Brown, Functional significance of the ipsilateral hemisphere during movement of the affected hand after stroke. Exp. Neurol. **190**, 425–432 (2004)

H. Shibasaki, M. Hallett, What is the Bereitschaftspotential? Clin. Neurophysiol. **117**, 2341–2356 (2006)

K. Stefan, E. Kunesch, L.G. Cohen, R. Benecke, J. Classen, Induction of plasticity in the human motor cortex by paired associative stimulation. Brain **123**(Pt 3), 572–584 (2000)

M. Stepien, J. Conradi, G. Waterstraat, F.U. Hohlefeld, G. Curio, V.V. Nikulin, Event-related desynchronization of sensorimotor EEG rhythms in hemiparetic patients with acute stroke. Neurosci. Lett. **488**(1), 17–21 (2010)

A.K. Thompson, F.R. Pomerantz, J.R. Wolpaw, Operant conditioning of a spinal reflex can improve locomotion after spinal cord injury in humans. J. Neurosci. **33**, 2365–2375 (2013)

G. Townsend, B. Grainmann, G. Pfurtscheller, Continuous EEG classification during motor imagery-simulation of an asynchronous BCI. IEEE Trans. Neural Syst. Rehabil. Eng. **12**, 258–265 (2004)

B. Várkuti, C. Guan, Y. Pan, K.S. Phua, K.K. Ang, C.W.K. Kuah, K. Chua, B.T. Ang, N. Birbaumer, R. Sitaram, Resting state changes in functional connectivity correlate with movement recovery for BCI and robot-assisted upper-extremity training after stroke. Neurorehabil. Neural Repair **27**, 53–62 (2013)

R. Xu, N. Jiang, N. Mrachacz-Kersting, C. Lin, G. Asin, J. Moreno, J. Pons, K. Dremstrup, D. Farina, A closed-loop brain-computer interface triggering an active ankle-foot orthosis for inducing cortical neural plasticity. IEEE Trans. Biomed. Eng. **9294**, 1 (2014a)

R. Xu, N. Jiang, A. Vuckovic, M. Hasan, N. Mrachacz-Kersting, D. Allan, M. Fraser, B. Nasseroleslami, B. Conway, K. Dremstrup, D. Farina, Movement-related cortical potentials in paraplegic patients: abnormal patterns and considerations for BCI-rehabilitation. Front Neuroeng. **7**, 35 (2014b)

ALS Population Assessment of a Dynamic Stopping Algorithm Implementation for P300 Spellers

B. Mainsah, L. Collins, K. Colwell, E. Sellers, D. Ryan, K. Caves
and C. Throckmorton

Keywords Electroencephalography · P300 speller · Dynamic stopping · Language model

Introduction

P300-based BCI systems are an attractive option for restoring communication abilities in people with severe physical limitations due to their non-invasiveness, minimal training requirements and long-term signal stability (Moghimi et al. 2013; Mak et al. 2011). These systems rely on event-related potentials (ERPs) in electroencephalography (EEG) data as control signals to enable users to make selections from an array of choices or action-encoded icons (Farwell and Donchin 1988). The occurrence of a rarely occurring desired event within a random sequence of stimulus events elicits a distinct ERP response which includes a large positive deflection called the P300 signal (Sutton et al. 1965). The P300 speller has been shown to be a viable communication alternative, especially in target BCI end-users (Mak et al. 2011), such as those with amyotrophic lateral sclerosis (ALS), who may eventually lose the ability to control muscle-based communication devices (Sellers and Donchin 2006; Sellers et al. 2010; Ortner et al. 2011).

Although there are a few P300-based BCI systems available for commercial use, e.g. (IntendiX 2010), transitioning these systems for independent home-based communication still poses several challenges (Sellers et al. 2010; Berger 2008;

B. Mainsah · L. Collins (✉) · K. Colwell · C. Throckmorton
Department of Electrical and Computer Engineering, Duke University, Durham, USA
e-mail: leslie.collins@duke.edu

E. Sellers · D. Ryan
Department of Psychology, East Tennessee State University, Johnson City, USA

K. Caves
Departments of Surgery, Medicine and Biomedical Engineering, Duke University,
Durham, USA

© The Author(s) 2015
C. Guger et al. (eds.), *Brain-Computer Interface Research*,
SpringerBriefs in Electrical and Computer Engineering,
DOI 10.1007/978-3-319-25190-5_8

Brunner et al. 2011). One limitation is their relatively slow selection rates when compared to other assistive communication aids. EEG activity reflects the spatial summation of electrical potentials which are severely attenuated through bone and tissue by the time they reach the scalp. Consequently, EEG data is characterized by its low spatio-temporal resolution and the relatively low-amplitude elicited ERPs are embedded within very noisy EEG data (Heinrich and Bach 2008). To enhance the signal quality and improve system accuracy, EEG data is averaged over multiple measurements (i.e. multiple presentations of the target character) to increase the signal-to-noise ratio (SNR) of the elicited ERPs. The standard operational approach has been to collect and average a fixed amount of data prior to character selection, termed *static data collection* (or *static stopping*), and the amount of data is usually consistent across users. Most P300 speller studies in people with disabilities have been implemented using static data collection (Moghimi et al. 2013; Mak et al. 2011; Sellers and Donchin 2006; Sellers et al. 2010; Ortner et al. 2011; Spüler et al. 2012; Riccio et al. 2013; Hoffmann et al. 2008; Townsend et al. 2010). However, static data collection might not be the most effective data collection strategy due to acute changes in an individual's EEG signal quality (for example, due to fatigue), and this may impact classification accuracy, and consequently spelling performance.

In recent years, data collection strategies have shifted towards adaptive methods which vary the amount of data collection based on some threshold function, an approach termed *dynamic stopping*. A dynamic stopping algorithm allows the BCI system to adjust the amount of data collection based on acute changes in a user's SNR level. Adaptive data collection strategies can potentially result in BCI systems with selection rates that are more practical for communication, as they have been shown to improve performance when compared to the static data collection strategy (Serby et al. 2005; Lenhardt et al. 2008; Liu et al. 2010; Thomas et al. 2013; Haihong et al. 2008; Park et al. 2010; Speier et al. 2014; Orhan et al. 2012; Jin et al. 2011; Hohne et al. 2010; Schreuder et al. 2013). Additionally, as is the case with much of the BCI literature (Moghimi et al. 2013; Mak et al. 2011; Kübler et al. 2013), studies with dynamic stopping algorithms have focused on offline simulations and online studies in healthy participants. Time, expense and logistical considerations are usually limiting factors in developing studies using people with disabilities. Nonetheless, online testing in target end-users is a key step in BCI algorithm development because algorithms optimized for healthy individuals may not generalize to the target population due to more variability in disease cause or disability progression.

Therefore, we focus on all three steps during BCI algorithm development: offline analyses for proof of concept, online testing/validation in healthy participants and online testing/validation in individuals with disabilities. We have developed a probabilistic algorithm that performs flash-to-flash analysis of EEG data to dynamically vary the amount of data collection prior to character selection. The system calculates a confidence level after each stimulus presentation (or flash) and data collection is stopped when the specified confidence level is achieved (Throckmorton et al. 2013). This approach has several advantages. By utilizing a probabilistic approach to select the target character, the algorithm automatically collects more data under low SNR conditions and less data under high SNR conditions without assuming a fixed level of user performance. Further, a probabilistic system readily allows adaptation to include additional knowledge, e.g. *a priori* information about the user's language (Mainsah et al. 2013), to inform the algorithm's behavior without having to redesign the system. In online studies in healthy participants, we have demonstrated statistically significant improvements in user performance using our dynamic stopping algorithms (Throckmorton et al. 2013; Mainsah et al. 2013). In this study, we validate the dynamic data collection algorithms in individuals with ALS, by comparing P300 speller performance with our algorithm to that using static data collection.

Methods

Ten participants with ALS (see Table 1) were recruited from across North Carolina and Tennessee by the Collins' Engineering Laboratory at Duke University and the Sellers' Laboratory at East Tennessee State University, with all experimental protocols approved by the respective university's institutional review boards. The open source BCI2000 software (Schalk et al. 2001) was used to implement the P300 speller with additional functionality for the dynamic data collection algorithms. The color checkerboard paradigm was used for stimulus presentation on a 6 × 6 alphanumeric grid of characters (Ryan et al. 2013). Data collected from electrodes Fz, Cz, P3, Pz, P4, PO7, PO8, and Oz were used for signal processing, with feature extraction and classifier training performed according to (Krusienski et al. 2008).

Table 1 Demographic information for study participants

Participant	Age	Sex	ALSFRS-R score	Years post 1st symptoms
D05	60	M	32	5.5
D06	59	M	3	7
D07	59	F	21	7
D08	62	F	42	11
E03	44	F	21	16
E20	38	M	5	8
E21	63	F	1	8
E23	57	M	30	2
E24	56	M	33	3
E25	49	M	33	1

Participants with "D" in their participant identifier were recruited at Duke University and those with "E" were recruited at East Tennessee State University. ALSFRS-R denotes the "ALS Functional Rating Scale", which provides a physician-generated estimate of the patient's degree of functional impairment, on a scale of 0 (high impairment) to 48 (low impairment) (Cedarbaum et al. 1999)

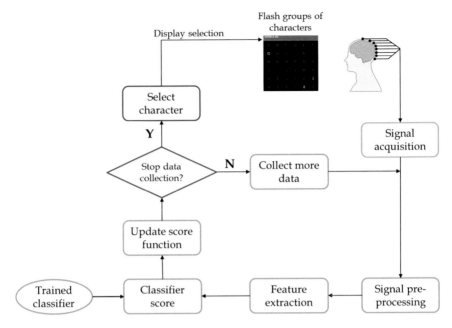

Fig. 1 P300 Speller components. The user is presented with a grid of character choices in an on-screen array. External electrodes are used to measure electroencephalography (EEG) signals from the scalp. The EEG signals are amplified, filtered and digitized for signal processing. Time sample blocks following flashes are used to extract feature vectors, which are scored with a trained classifier, and the cumulative character scores are updated following each flash. After a certain amount of data collection, the character with the largest classifier response is selected and presented as the user's intended choice

Participants performed word copy spelling tasks during a BCI session, with feedback, with the color checkerboard paradigm for stimulus presentation. Figure 1 shows the basic operation of a P300 speller. EEG data is acquired from a select number of electrodes on the scalp. To copy-spell a character, a user focuses on a desired (or target) character as groups of characters are flashed on a screen: the illumination of the relatively rare target character elicits a P300 ERP response. After a subset of characters is flashed on the screen, features are extracted from a time-window of EEG data and the user-specific trained classifier is used to score the features. This classifier score is used to update the cumulative classifier scores of all of the grid characters according to the system's update rules. After a certain amount of data collection, the character with the highest cumulative classifier score is selected by the system as the user's intended target.

The amount of data collection prior to character selection can be fixed or dynamically determined. In the 6 × 6 checkerboard paradigm, there are 18 flashes/sequence with each character flashed twice in a sequence. A maximum limit of 7 sequences was set for all data collection algorithms across all users. The data collection algorithms used in this study are described below:

Static Stopping (SS): In the SS algorithm, the amount of data collected was fixed. Prior to data collection, all characters in the grid are assigned a score of zero. With each flash, only the scores of each character in the flashed subset are updated by adding the current classifier score. After data collection, the character with the maximum score is selected as the target character.

Dynamic Stopping (DS): In the DS algorithm, a probability distribution is maintained over all characters in the grid and this probability represents the level of confidence that a character is the target character (see Fig. 2). Prior to data collection, all characters are initialized with a uniform probability of being the target. With each flash, the probability values are updated with new information from the EEG data via each classifier score that is integrated into the model via a Bayesian update process (Throckmorton et al. 2013). Data collection is stopped when a particular character's probability exceeds a threshold value (in this study a threshold value of 0.90 was used) and that character is selected as the target character.

Dynamic Stopping with Language Model (DSLM): The DS algorithm is enhanced by incorporating information about the English language via a statistical language model, which captures the predictability of the ordering of letters in words (Mainsah et al. 2013). For example, vowels are more likely to follow consonants, and specifically, a "U" is most likely to follow a "Q". The DSLM algorithm Bayesian update process is identical to that of the DS algorithm except for the

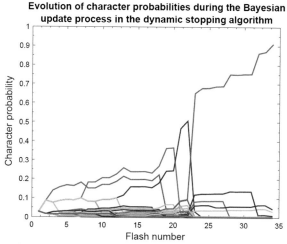

Fig. 2 Evolution of character probabilities in the dynamic stopping algorithm. Prior to data collection, character probabilities are initialized uniformly. This figure shows the evolution of the character probabilities for all of the characters as a function of the flash number. Following each flash, probability values are updated via the Bayesian update process by incorporating the current classifier score. After several flashes, there is separation between likely and unlikely characters and this separation increases with more data collection. Eventually, the probability of one character converges to 1 and ideally this should correspond to the target character. Data collection is stopped when a character's probability attains a preset threshold value and that character is selected by the BCI system as the user's intended target character

initialization probabilities. The character probabilities are initialized based on the previous character selection (a bigram model developed from The CMU Pronouncing Dictionary 2013), with an error factor incorporated to account for the possibility of character misspellings.

Each P300 speller session consisted of a training phase to collect data for classifier training, and a testing phase for the copy-spelling task, with the order of the three algorithms randomized during each session. Following each session, participants were asked subjective questions to assess their preference among the three data collection algorithms.

Results

Participants were evaluated on the following performance measures, pooled across all three sessions: task completion time, accuracy, and bit rate (McFarland et al. 2003). Figure 3a shows the average time it took to spell each character, as well as the total spelling task completion (36 characters). Compared to static data collection

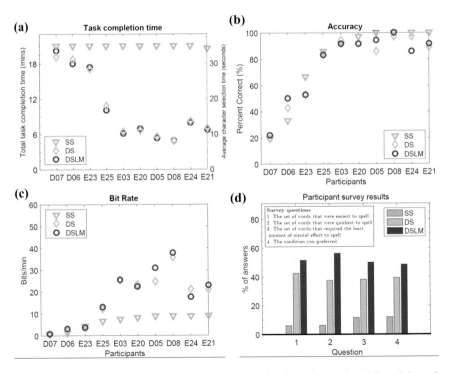

Fig. 3 Comparison of performance measures between static, dynamic stopping (DS) and dynamic stopping with language model (DSLM) for **a** Task completion time, **b** Accuracy, **c** Bit Rate, and **d** Post-session preference survey

at 35 s per character selection, participants had a significant decrease in character selection time (45–75 %), mostly ranging from 8 to 20 s/character selection (p < 0.00001). The results show a significant reduction in the amount of data collected by the dynamic stopping algorithms and, importantly, accuracy levels did not significantly deteriorate from static data collection levels, as shown in Fig. 3b (p < 0.23). The combination of maintaining similar accuracy levels while significantly reducing data collection resulted in a significant improvement in bit rate from static (6.44 ± 3.21 bits/min) to DS (17.06 ± 11.78 bits/min) and DSLM (17.82 ± 12.54 bits/min), p < 0.00001, Fig. 3c. Based on the post-session survey results, shown in Fig. 3d, despite varying accuracy levels, participants overwhelmingly preferred the dynamic data collection algorithms.

Discussion

Our dynamic stopping algorithms assess EEG data to maximize the spelling speed of the P300 speller without compromising accuracy. The amount of data collected per character varied across and within participants, demonstrating the capability of the algorithm to adapt to acute changes in user performance rather than relying on the collection of a fixed amount of data. Most participants experienced a significant reduction in task completion times with the dynamic data collection algorithms with little to no negative impact on accuracy. Further, communication rates were greatly improved for the majority of participants. Despite a wide range in accuracy levels, participants predominantly preferred the dynamic data collection algorithms.

Validation in target end users is a key step in BCI algorithm development. Algorithm improvements demonstrated in users without disability (a generally uniform population which skews young) may not necessarily hold true for users with disability who exhibit a wider variability due to disease cause or progression. The DSLM condition was statistically equivalent to the DS condition, which may be a function of small sample size because a study including a larger number of able-bodied participants showed that the DSLM condition resulted in significantly higher performance than the DS condition. In addition, the language model used in this study was a simple bigram model, hence only a limited amount of the user's spelling history was used to inform the character probability initialization process. There is the potential to improve performance with more complex language models, but it is important to test this in the target population.

The needs of BCI target end-users have to factor in the design process, and these considerations may not be obvious if making inferences from individuals without disability. An algorithm or strategy might have the potential to improve performance, but may be integrated in a way that it increases a user's cognitive load or discomfort. For example, predictive word options are ubiquitous in most communication devices to increase throughput. However, if not properly incorporated into a BCI system, it can lead to diminished ERP responses due to the additional attentional task of processing new information via a changing interface. The language information used in this study was integrated in the background, resulting in no change in the interface. The participants responded positively to the dynamic data collection algorithms, indicating the desire to achieve faster BCI selection rates. To improve performance, complex language models can be integrated in a similar manner or within the interface in a manner that doesn't lead to increased cognitive load.

This study evaluated a viable BCI algorithm that has been validated in people with disabilities with results indicating the potential advantage of using an adaptive data collection strategy to improve P300 speller selection rates. Further development includes evaluation with sentence spelling tasks, where there is the potential to further enhance performance using natural language processing tools such as word prediction and/or dictionary-based spelling correction.

Acknowledgments This research project was funded by NIH/NIDCD under grant number R33 DC010470.

References

T.W. Berger, *Brain-Computer Interfaces: An International Assessment of Research and Development Trends* (Springer, 2008)

P. Brunner, L. Bianchi, C. Guger, F. Cincotti, G. Schalk, Current trends in hardware and software for brain–computer interfaces (BCIs). J. Neural Eng. **8**, 025001 (2011)

J.M. Cedarbaum et al., The ALSFRS-R: a revised ALS functional rating scale that incorporates assessments of respiratory function. J. Neurol. Sci. **169**, 13 (1999)

L.A. Farwell, E. Donchin, Talking off the top of your head: toward a mental prosthesis utilizing event-related brain potentials. Electroencephalogr. Clin. Neurophysiol. **70**, 510 (1988)

Z. Haihong, G. Cuntai, W. Chuanchu, Asynchronous P300-based brain-computer interfaces: a computational approach with statistical models. IEEE Trans. Biomed. Eng. **55**, 1754 (2008)

S. Heinrich, M. Bach, Signal and noise in P300 recordings to visual stimuli. Doc. Ophthalmol. **117**, 73 (2008)

U. Hoffmann, J.-M. Vesin, T. Ebrahimi, K. Diserens, An efficient P300-based brain–computer interface for disabled subjects. J. Neurosci. Methods **167**, 115 (2008)

J. Hohne, M. Schreuder, B. Blankertz, M. Tangermann, in *2010 Annual International Conference of the IEEE Engineering in Medicine and Biology Society* (*EMBC*) (2010), pp. 4185–4188

IntendiX by g.tec, World's first Personal BCI Speller (2010)

J. Jin et al., An adaptive P300-based control system. J. Neural Eng. **8**, 036006 (2011)

D.J. Krusienski, E.W. Sellers, D.J. McFarland, T.M. Vaughan, J.R. Wolpaw, Toward enhanced P300 speller performance. J. Neurosci. Methods **167**, 15 (2008)

A. Kübler, E. Holz, T. Kaufmann, in *Converging Clinical and Engineering Research on Neurorehabilitation*, eds. by J.L. Pons, D. Torricelli, M. Pajaro (Springer, 2013), vol. 1, pp. 1271–1274

A. Lenhardt, M. Kaper, H.J. Ritter, An adaptive P300-based online brain computer interface. IEEE Trans. Neural Syst. Rehabil. Eng. **16**, 121 (2008)

T. Liu, L. Goldberg, S. Gao, B. Hong, An online brain–computer interface using non-flashing visual evoked potentials. J. Neural Eng. **7**, 036003 (2010)

J.N. Mak et al., Optimizing the P300-based brain–computer interface: current status, limitations and future directions. J. Neural Eng. **8**, 025003 (2011)

B. Mainsah, K. Colwell, L. Collins, C. Throckmorton, Utilizing a Language model to improve online dynamic data collection in P300 spellers. IEEE Trans. Neural Syst. Rehabil. Eng.: Publ. IEEE Eng. Med. Biol. Soc. (2013)

D.J. McFarland, W.A. Sarnacki, J.R. Wolpaw, Brain-computer interface (BCI) operation: optimizing information transfer rates. Biol. Psychol. **63**, 237 (2003)

S. Moghimi, A. Kushki, A.M. Guerguerian, T. Chau, A review of EEG-based brain-computer interfaces as access pathways for individuals with severe disabilities. Assistive Technol. Official J. RESNA **25**, 99 (2013) (Summer)

R. Ortner et al., in *2011 IEEE Symposium on Computational Intelligence, Cognitive Algorithms, Mind, and Brain* (*CCMB*) (2011), pp. 1–6

U. Orhan et al., in *2012 IEEE International Conference on Acoustics, Speech and Signal Processing* (*ICASSP*) (2012), pp. 645–648

J. Park, K.-E. Kim, S. Jo, in *Proceedings of the 15th international conference on Intelligent user interfaces* (ACM, 2010), pp. 1–10

A. Riccio et al., Attention and P300-based BCI performance in people with amyotrophic lateral sclerosis. Front. Hum. Neurosci. **7** (2013)

D. Ryan, K. Colwell, C. Throckmorton, L. Collins, E. Sellers, *Paper Presented at the 5th International BCI Meeting* (Asilomar, CA, USA, 2013)

G. Schalk, D.J. McFarland, T. Hinterberger, N. Birbaumer, J.R. Wolpaw, BCI2000: development of a general purpose brain-computer interface (BCI) system. Soc. Neurosci. Abs. **27**, 168 (2001)

M. Schreuder et al., Optimizing event-related potential based brain–computer interfaces: a systematic evaluation of dynamic stopping methods. J. Neural Eng. **10**, 036025 (2013)

E.W. Sellers, E. Donchin, A P300-based brain-computer interface: initial tests by ALS patients. Clin. Neurophysiol. **117**, 538 (2006)

E.W. Sellers, D.J. McFarland, T.M. Vaughan, J.R. Wolpaw, in *Brain-Computer Interfaces* (Springer, 2010), pp. 97–111

H. Serby, E. Yom-Tov, G.F. Inbar, An improved P300-based brain-computer interface. IEEE Trans. Neural Syst. Rehabil. Eng. **13**, 89 (2005)

W. Speier, C. Arnold, J. Lu, A. Deshpande, N. Pouratian, Integrating language information with a hidden Markov model to improve communication rate in the P300 speller. IEEE Trans. Neural Syst. Rehabil. Eng. **22**, 678 (2014)

M. Spüler et al., Online use of error-related potentials in healthy users and people with severe motor impairment increases performance of a P300-BCI. Clin. Neurophysiol. **123**, 1328 (2012)

S. Sutton, M. Braren, J. Zubin, E.R. John, Evoked-potential correlates of stimulus uncertainty. Science **150**, 1187 (1965)

E. Thomas et al., in *2013 6th International IEEE/EMBS Conference on Neural Engineering* (*NER*) (2013), pp. 1062–1065

C.S. Throckmorton, K.A. Colwell, D.B. Ryan, E.W. Sellers, L.M. Collins, Bayesian approach to dynamically controlling data collection in P300 spellers. IEEE Trans. Neural Syst. Rehabil. Eng.: Publ. IEEE Eng. Med. Biol. Soc. **21**, 508 (2013)

G. Townsend et al., A novel P300-based brain-computer interface stimulus presentation paradigm: moving beyond rows and columns. Clin. Neurophysiol.: Official J. Int. Fed. Clin. Neurophysiol. **121**, 1109 (2010)

The CMU Pronouncing Dictionary (Carnegie Mellon University, 2013)

Semi-autonomous Hybrid Brain-Machine Interface

David P. McMullen, Matthew S. Fifer, Brock A. Wester, Guy Hotson,
Kapil D. Katyal, Matthew S. Johannes, Timothy G. McGee,
Andrew Harris, Alan D. Ravitz, Michael P. McLoughlin,
William S. Anderson, Nitish V. Thakor and Nathan E. Crone

Abstract Although advanced prosthetic limbs, such as the modular prosthetic limb (MPL), are now capable of mimicking the dexterity of human limbs, brain-machine interfaces (BMIs) are not yet able to take full advantage of their capabilities. To improve BMI control of the MPL, we are developing a semi-autonomous system, the Hybrid Augmented Reality Multimodal Operation Neural Integration Environment (HARMONIE). This system is designed to utilize novel control strategies including hybrid input (adding eye tracking to neural control), supervisory control (decoding high-level patient goals), and intelligent robotics (incorporating computer vision and route planning algorithms). Patients use eye gaze to indicate a desired object that has been recognized by computer vision. They then perform a desired action, such as reaching and grasping, which is decoded and carried out by the MPL via route planning algorithms. Here we present two patients, implanted with electrocorticography (ECoG) and depth electrodes, who controlled the HARMONIE system to perform reach and grasping tasks; in addition, one patient also used the HARMONIE system to simulate self-feeding. This work

---

D.P. McMullen and M.S. Fifer are co-first authors.

D.P. McMullen (✉) · W.S. Anderson
Department of Neurosurgery, Johns Hopkins University,
600 N Wolfe Street/Meyer 8-181, 21287 Baltimore, MD, USA
e-mail: dmcmull4@jhmi.edu

M.S. Fifer
Department of Biomedical Engineering, Johns Hopkins University, Baltimore, USA

B.A. Wester · K.D. Katyal · M.S. Johannes · T.G. McGee · A. Harris · A.D. Ravitz
M.P. McLoughlin
Applied Physics Laboratory, Johns Hopkins University, Baltimore, USA

G. Hotson · N.V. Thakor
Department of Electrical Engineering, Johns Hopkins University, Baltimore, USA

N.E. Crone
Department of Neurology, Johns Hopkins University, Baltimore, USA

© The Author(s) 2015
C. Guger et al. (eds.), *Brain-Computer Interface Research*,
SpringerBriefs in Electrical and Computer Engineering,
DOI 10.1007/978-3-319-25190-5_9

builds upon prior research to demonstrate the feasibility of using novel control strategies to enable patients to perform a wider variety of activities of daily living (ADLs).

Keywords Brain-machine interface (BMI) · Brain-computer interface (BCI) · Electrocorticography (ECoG) · Hybrid BCI · Supervisory control · Intelligent robotics

Introduction

Despite recent advances in robotic prosthetic arms, neuroprosthetics powered by brain-machine interfaces (BMIs) have not experienced widespread clinical adoption. Neural control of prosthetic limbs has traditionally required the patient to micromanage a multitude of joints in the arm and hand simultaneously. This method, however, can only provide control over a fraction of what advanced prosthetic limbs, such as the Modular Prosthetic Limb (MPL) developed by the Johns Hopkins University Applied Physics Laboratory (APL), can accomplish.

By implementing novel control strategies, such as hybrid BMI, supervisory control, and intelligent robotics, BMIs can be designed to offload some of the mental burden of neuroprosthetic control (Allison et al. 2012). Hybrid BMIs augment primary neural signal recordings with various other sensing modalities to maximize the number of usable signals that can be obtained from a patient (Allison et al. 2010; Pfurtscheller et al. 2010). These recordings can combine multiple types of neural signals, such as combining single-unit recordings and LFPs in a multi-electrode array (MEA)-based BMI or combining SSVEP and P300 potentials in an EEG-based BMI (Millán et al. 2010; Pfurtscheller et al. 2010). These neural hybrid BMIs make full use of the wide scope of neural signals that have been characterized for control. Beyond recording additional neural signals, physiological hybrid BMIs can take advantage of other functioning patient anatomy to record electrooculographic (EOG) signals or eye tracking. Eye tracking is particularly intriguing as it allows the BMI to take advantage of an intuitive physiological signal, that of gaze end-point, and translate the natural intention to interact with a desired object into use by the BMI (Zander et al. 2010; Lee et al. 2010; Frisoli et al. 2012; Kim et al. 2014).

Beyond maximizing the control power of the BMI by using additional signals, hybrid BMIs are notable for their ability to limit the number of false positives (Pfurtscheller et al. 2010). Multiple independent control modalities decrease the chances of misclassifications occurring in all modalities at once. Additionally, error potentials might also be recorded to indicate BMI movements aren't what were intended by the user (Milekovic et al. 2013). The ability to limit false positives has practical applications in semi-autonomous assistive BMIs, where inadvertent initiation of a task can lead to a lengthy and unwanted robotic actuation.

While hybrid BMIs maximize the number and type of signals recorded from a patient, supervisory control strategies focus on what high-level goals can be decoded from these signals. Supervisory control signals have been explored as part of cognitive BMIs (Musallam et al. 2004; Andersen et al. 2004; Hayati and Venkataraman 1989; Flemisch et al. 2003; Goodrich et al. 2006; Wolpaw 2007). These signals separate the intent of the action (i.e., intending to pick an object up) from the low-level motor plan for carrying out the action (i.e., controlling the kinematics of joint angles and grasp types). Decoding a goal can offer performance advantages, especially if the goal signal appears before the movement starts, as opposed to during the execution of a movement. Removing the need for continuous low-level control of prosthetics by their users would offload some of the cognitive burden and daily re-training required to perform activities of daily living (ADLs), which are a major focus of decoding high level signals (Royer and He 2009; Royer et al. 2011).

Supervisory control can only be implemented if there is a modality that can rapidly capture the neural correlates of goal intent. ECoG and depth electrodes, in particular, are emerging recording modalities for BMIs. These electrodes record from the cortical surface electrodes or from depth electrodes spanning cortical and subcortical structures. ECoG provides an intermediate spatial scale, sampling from a much smaller area than scalp electroencephalography (EEG) while providing much wider area coverage than microelectrode arrays (MEAs). Relative to scalp EEG, ECoG also provides greater access to high gamma activity ($\sim$70–150 Hz), which has been demonstrated to reflect the spiking activity of local neuronal populations. High gamma activity has been validated as a reliable indicator of cortical processing across functional domains, exhibiting activation during movements (Crone et al. 1998), selective attention (Ray et al. 2008), visual processing (Lachaux et al. 2005), auditory processing (Crone et al. 2001), and speech (Crone et al. 2001). Furthermore, high gamma activity has greater spatial and temporal resolution than alpha (8–12 Hz) and beta (16–30 Hz) activity (Crone et al. 2011). These properties make high gamma activity from ECoG an attractive candidate for use as a control signal in a BMI. To date, high gamma activity from ECoG electrodes has been used by our team and others for online control of computer cursors (Leuthardt et al. 2004; Schalk et al. 2008; Wang et al. 2013) and prosthetic arms performing reaches (Wang et al. 2013; Yanagisawa et al. 2012) or grasps (Wang et al. 2013; Yanagisawa et al. 2012; Vinjamuri et al. 2011) controlled separately, or reaches and grasps controlled simultaneously (Fifer et al. 2014).

Hybrid BMIs and supervisory control strategies allow researchers to take full advantage of advanced prosthetic limbs, such as the APL's MPL. To interact with the outside world in a naturalistic way, a paralyzed BMI user must have access to an end effector that is capable of replicating or approximating movements of the arm and hand. As a lead site for the DARPA Revolutionizing Prosthetics Program, the JHU/APL has developed the Modular Prosthetic Limb (MPL), a state-of-the-art anthropomorphic robotic upper limb. The MPL is roughly the weight of a human arm and has been designed to accommodate patients with varying levels of amputation. The MPL is controlled by an extremely flexible software interface that can either route commands to the physical MPL or its virtualization in the Virtual Integration Environment (VIE), a customizable simulation of control over the MPL and its interaction with virtual objects.

Intelligent robotics provides the framework for incorporating hybrid and supervisory control signals into a controller for the MPL that is usable by paralyzed patients. Intelligent robotics incorporate environmental information to adapt control of a robotic limb in a semi-autonomous fashion for object recognition, pathway optimization and obstacle avoidance (Iturrate et al. 2009; Escolano et al. 2012). For example, an intelligent BMI system might allow the patient to identify an object of interest via eye control and indicate a high-level goal (e.g., reach and grasp an object) to trigger the MPL to carry out a task (Fig. 1). By increasing the complexity, accuracy, and speed of MPL movements, intelligent robotics will allow patients to use their prosthetic limbs for longer periods of time and in ways not previously possible in their daily lives.

To date, patients faced with significant motor and communication difficulties have had limited treatment options. We have developed an intuitive motor BMI system, the Hybrid Augmented Reality Multimodal Operation Neural Integration Environment (HARMONIE), to allow patients with a wide range of disabilities to gain greater autonomy and functionality in their daily lives. We anticipate that this could lead to decreased therapy costs, reduced burden on caregivers, and improved mental health for the patients themselves. Here we detail our initial demonstrations

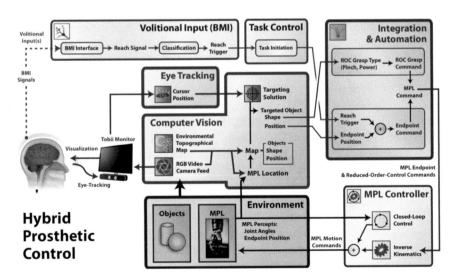

Fig. 1 Block diagram of the HARMONIE BMI system. Hybrid inputs are incorporated through eye tracking of objects on the monitor. BMI signals decode high-level goals. Computer vision segments the objects in the environment and the intelligent robotic system integrates this information to carry out the motor task (figure courtesy of IEEE TNSRE McMullen et al. 2014)

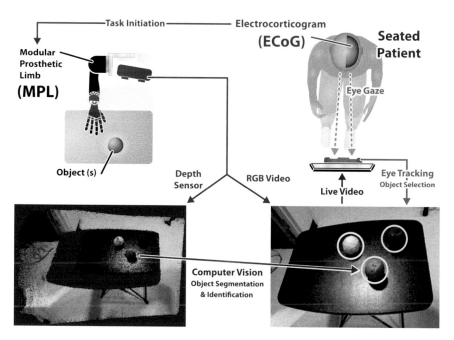

Fig. 2 Schematic of the HARMONIE experimental setup in studies with human epilepsy patients. The patient views a segmented scene sent from the computer vision device. The Kinect sensor records both depth information and RGB video of the experimental workspace. The patient's gaze is recorded by monitor-mounted eye tracking and movement intention is detected by the neural interface. The MPL carries out the complex motor task involving reaching, grasping, and manipulation of an object (figure courtesy of IEEE TNSRE McMullen et al. 2014)

of the HARMONIE system using hybrid input, supervisory control, and intelligent robotics with epilepsy patients implanted with ECoG electrodes (Fig. 2) (McMullen et al. 2014; Katyal et al. 2013).

Methods

Spatio-Temporal Functional Mapping (STFM)

Neural feature selection for control of the MPL and the HARMONIE system was performed by our team using a custom platform designed in MATLAB (MathWorks; Natick, MA). Our STFM software (Fig. 3) rapidly assesses spatio-temporal modulation of task-related ECoG high gamma activity relative to baseline (McMullen et al. 2014; Fifer et al. 2014). ECoG signals and behavioral trigger information are streamed into an experimental workstation. The STFM platform uses the timing of the behavioral triggers to segment ECoG signals into pre-cue baseline periods and post-cue analysis periods. High gamma activity is

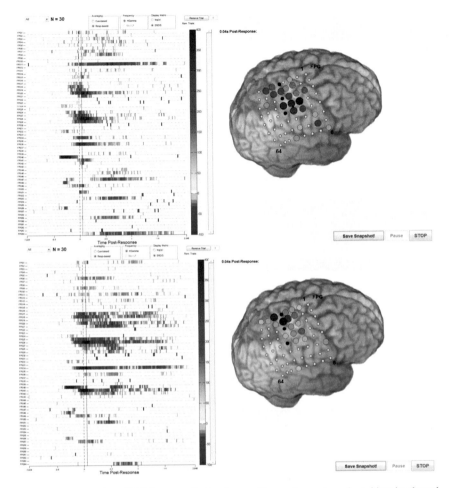

Fig. 3 Screenshots of the STFM system display for reaching and grasping of a subject implanted with ECoG electrodes. The *vertical axis* of each raster represents distinct channels, while the *horizontal axis* represents time after the onset of the movement, denoted by the *dotted line*. On the *right side* of each screenshot is a spatial map of the neural activity at the time point denoted by the *solid line* in the raster (i.e., 0.04 s post-movement-onset). The top screenshot is a raster and map of grasping, while the bottom screenshot depicts the pattern of activity for forward reaching

extracted from both periods in 128 ms windows, sliding by 16 ms, using a Hilbert transform with embedded 70–120 Hz bandpass filter. Statistical thresholding is performed by comparing each post-cue window to the distribution of baseline high gamma activity in the corresponding channel. False discovery rate (FDR) correction is used to limit the number of false positives due to performing multiple tests (Benjamini and Hochberg 1995). The power in each significantly modulated post-cue window is then displayed in a channel x time raster. An additional layer of analysis can be performed where post-cue windows can be compared across movement conditions (e.g., reach vs. grasp) rather than between periods of activity and baseline. Together, the maps of task-relevance and task-specificity generated by the STFM platform provide an excellent source of information about which electrodes are suitable for controlling reaching and/or grasping of the MPL.

Brain-Machine Interface (BMI) Decoding Models

Neural features were selected as inputs to parallel binary classifiers for reaching and grasping. For the efforts detailed in this chapter, the features chosen were high gamma (70–120 Hz) power recorded at individual ECoG sites in 400 ms windows using a 16th order Burg autoregressive spectral estimate. Linear discriminant analysis (LDA) models were used to decode whether the patient was actively reaching and/or actively grasping in each window. For the second subject in this study, a slight improvement was made to the decoding model structure. Transition probabilities were embedded in the model as prior probabilities to fine-tune the preference for staying in or leaving the current state—reaching versus not reaching and grasping versus not grasping. These probabilities greatly increased the smoothness of the movements generated by the neural decoding models. These models were trained in offline sessions where the patient was verbally cued to either "Reach," "Grasp," or "Reach and Grasp." The second subject for the study described below in results (McMullen et al. 2014) received an intermediate training session where the outputs from decoding models trained offline were used to actuate a virtualization of the MPL in the a virtual framework described below. Decoding models were trained on the recordings from this session and subsequently used for control of the physical MPL. To initiate reach, grasp, and place movements, only the output from the reaching versus not reaching decoder was used. Self-feeding trials required the use of both decoders at various stages.

Modular Prosthetic Limb (MPL)

The MPL has 17 controllable degrees of freedom (DoF) and 26 articulating DoF. The MPL is also highly sensorized, with 190 sensors to monitor the angle, angular velocity, and torque of each joint, including 10 contact sensors each in the fingertips and palm (Fig. 4). The MPL is controlled via UDP commands through the VulcanX software interface that allows for direct control over joint angles, joint angular velocities, three-dimensional hand position, and/or progression through preprogrammed grasps. VulcanX can alternatively (or simultaneously) send command signals to the virtual MPL (vMPL) in the Virtual Integration Environment (VIE), described in more detail below. In two studies to date (McMullen et al. 2014; Fifer et al. 2014), our team has used 3-D endpoint (EP) velocity control and reduced order control (ROC) of preprogrammed grasps.

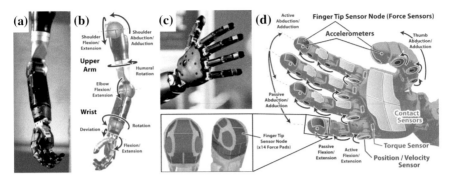

Fig. 4 Representations of the MPL. Panel **b** demonstrates the range of motion capabilities of the shoulder, arm, and wrist of the MPL (greater detail in panel **a** photograph). Panel **d** shows both the range of motion of the hand and its sensing capabilities. As demonstrated in the *inset* of panel **d**, the MPL hand (pictured in panel **c**) has 14 force sensor pads on each fingertip node

Onboard the MPL is a hierarchical hardware-software architecture for coordinating the movements of the shoulder, wrist, and finger joints. The Neural Fusion Unit is a processor capable of implementing neural decoding algorithms and generating limb motion commands for driving the MPL. A Limb Controller receives limb motion commands and coordinates the actuation of the 17 individual motor controllers. Ten Small Motor Controllers (i.e., for the fingers), four Large Motor Controllers (i.e., for the upper arm), and three Wrist Motor Controllers receive data from temperature, torque, and position sensors in real-time for use in closed-loop positioning of individual joints. Information is relayed to and from the MPL via CAN bus.

Virtual Integration Environment (VIE)

The VIE is a software emulation of the MPL with shared communication interfaces, designed as an integration and training tool for clinical research in the fields of upper extremity prosthetics and rehabilitation (Fig. 5). The VIE provides both a 3D graphical visualization and physical simulation of the MPL, including the 26 articulating joints and 17 independently controllable virtual motors operating and interacting within a virtual world. The VIE simulates physical object interactions with the virtual MPL (vMPL), including contact, grasping and fingertip force, and allows grasping, transporting, and repositioning of objects. Object interactions governed by the physics engine within the virtual environment result in calculated forces based on defined object properties (e.g. mass, material properties). Each of the 14 pads of the virtual Finger Tip Sensor Nodes (FTSN) in the vMPL finger and thumb tips report a dynamic force as normally reported by a physical capacitive plate force sensor in the physical MPL. This force considers the sum of the torques

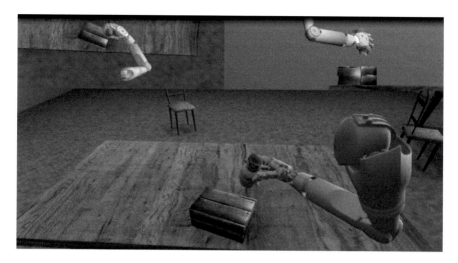

Fig. 5 vMPL within the VIE. A virtual MPL is shown within the 3-D virtual integration environment. This allows users to interact with the limb from any angle, as in real life. The virtual nature of the system allows rapid incorporation of robotic advancements and a nearly limitless range of environmental scenarios for the patient to interact with, including standard rehabilitation metrics

placed on each limb segment proximal to the object collision, the location of the collision on the fingertip with respect to the sensor pad footprint, and the direction and normality of the collision respective to the sensor pad orientation. These forces, in addition to limb data and virtual sensor percepts, are generated through regular control of the vMPL and event handling within the environment. Via defined communication interfaces, these data and percepts can be streamed to the user via UDP. The VIE has been developed as a framework allowing creation of multiple virtual scenarios that facilitate clinical training and real-time operation of the vMPL, as well as offline data analysis. Additional scenarios have been designed to allow testing of basic rehabilitation measures.

Hybrid Augmented Reality Multimodal Operation Neural Integration Environment (HARMONIE)

The HARMONIE system is a prosthetics control framework enabling the utilization of both neurophysiological (e.g., ECoG signals or spike firing rates) and non-neural (e.g., eye-tracking) control signals. Sizes, shapes, 3-D orientations, and 3-D positions in the workspace are segmented from point cloud data collected by the depth sensor of a Microsoft Kinect (Redmond, WA, USA). The Kinect is mounted on the same stand as the MPL, and is directed down at a table directly in front of the MPL. Computer vision algorithms leveraging the open project Point Cloud Library

(a)   (b)

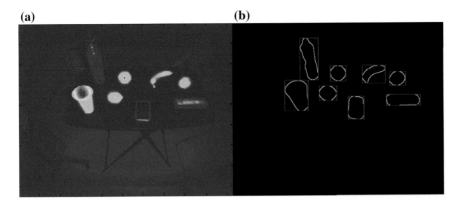

Fig. 6 Object segmentation with the Microsoft Kinect and trajectory modeling. **a** An infrared image of an example workspace populated with various objects. **b** Object identification. Object parameters (e.g., location, orientation, size) can then be extracted

(PCL) have been implemented in the Robotic Operating System (ROS) Fuerte platform (Willow Garage; Menlo Park, CA, USA) to localize and parameterize objects in the workspace (Fig. 6).

Once the target object properties are known, appropriate hand position and grasp configuration are computed. Inverse kinematic algorithms are used to translate hand position into a full arm posture. Path-planning algorithms compute a set of way-points to control the angle of approach to the target object. Sensors on the fingers and palm allow confirmation of grasp locations and desired forces. In studies to date, neural control has been integrated as a binary on/off switch in concert with eye-tracking that initiates or ceases movement in the desired trajectory (McMullen et al. 2014; Katyal et al. 2013). Eye-tracking was performed by a Tobii PCEye while patients viewed a red-green-blue (RGB) video stream from the Kinect.

Results

The HARMONIE system demonstrated here leveraged the approach used in a previous successful demonstration of independent neural control of reach and grasp (Fifer et al. 2014) to allow two subjects to perform ADLs. Both patients participated in the reach, grasp, and place task (Fig. 7), while the second patient also participated in the self-feeding exploratory study (Fig. 8).

Subject 1 attempted 28 self-paced trials and Subject 2 attempted 31 cued trials (Table 1). Success of the entire system from task intention to successful MPL actuation (termed a "global success") was analyzed, as were the success of ECoG decoding, computer vision, and MPL actuation components. Subject 1 successfully completed 20 of 28 trials (71.4 %), of which four failures were due to ECoG misclassification and three failed due to eye tracking. Subject 2 was successful in 21

Fig. 7 Demonstration of the HARMONIE system controlling reach and grasp. Computer vision segments the three spheres (*top left*) which the patient selected via eye control (*top middle*). The patient initiates the system with a reach to the desired target (*top right*). The MPL then performs the reach-grasp-and-drop motor task autonomously (*bottom panels*)

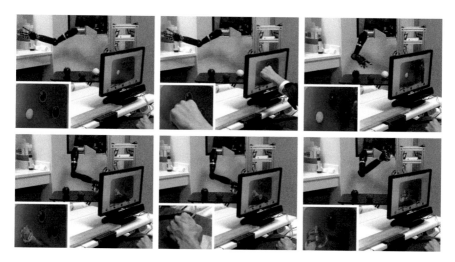

Fig. 8 Demonstration of the HARMONIE system controlling self-feeding. Computer vision segments the three spheres (*top left*) which the patient selected via eye control and then initiated with a reach (*top middle*). Once the MPL grasps the objects (*bottom left*), the patient initiates the self-feeding portion with a grasp (*bottom middle*), which leads the MPL to perform the action (*bottom right*)

Table 1 Results from a demonstration of the HARMONIE BMI system with two patients

Subject	# of electrodes	Global success rate (%)	ECoG-based balanced accuracy (* = p < 0.05) (%)	# of global false positives	Completion time (s)
1	3	71.4	91.1	1	22.3
2	4	67.7	92.9	0	12.2

of 31 trials (67.7 %), albeit with a harder task of choosing between three objects with which the MPL could interact. Nine of the failures were attributed to the route planning module, while only one was due to ECoG misclassification. There were a minimal number of false positives, defined as an undesired triggering of the MPL to interact with an object. Subject 1 experienced only one false positive and Subject 2 experienced none. This is in comparison to fourteen ECoG-only based false positives for Subject 1 and three for Subject 2 that would have occurred if neural decoding was the only deciding input into the BMI system.

The robotic and computer vision components of the HARMONIE system performed extremely well. The MPL successfully performed its portion of the task, once initiated, in 100 % of trials for Subject 1 and 70 % of trials for Subject 2. Improvements in movement transitions decreased the MPL's completion time from 22.3 s in Subject 1 to 12.2 s in Subject 2. The computer vision component of the system demonstrated robustness throughout the trials, and in offline testing, of its ability to detect and differentiate objects. The system simultaneously detected eight objects in an offline test workspace, with radii ranging from 3.35 to 2 cm. Objects could be successfully detected unless they were placed within 2.5 cm of each other.

In an additional session with Subject 2, we also demonstrated control of simulated self-feeding by the HARMONIE system using independent neural decoders for reaching and grasping. Due to time constraints, only nine such trials were attempted. A custom graphical user interface navigated by eye-controlled cursor and blink clicks enabled the subject to select whether the MPL should: (a) reach to and grasp the object in the workspace, (b) return to the home position, or (c) bring the object to the "face," a spot above and to the left of the shoulder mount. Once a mode was selected, the movement was initiated by the neural decoding model. A decoded reach was necessary to reach to and grab the object (Fig. 8, top middle panel); after the object was grasped, a decoded grasp (Fig. 8, bottom middle panel) would bring the object to the mouth (Fig. 8, bottom right panel), while a decoded reach would bring the object to the home position. The computer vision module accurately recognized the object in all nine trials. The neural decoding model successfully detected the intent to initiate a movement in eight of the trials. In six of nine trials, the subject was successfully able to completely simulate the entirely of self-feeding with an object in the workspace with a range of completion times from 15 to 25 s.

Discussion

The results demonstrate the advantage of the HARMONIE system, which uses novel strategies for controlling an intelligent BMI. The use of eye movements as an additional input modality provided an intuitive means of indicating the object of interest. Eye-tracking is a particularly attractive technology, since it is already commercially available and usable by the majority of paralyzed patients (i.e., those that are not completely locked-in). As mentioned above, hybrid inputs also greatly reduced the number of false positive initiations of the MPL. This is of particular benefit in semi-autonomous systems where there is a high time cost for proceeding with the entire motor plan if incorrectly triggered. Future iterations of the HARMONIE system may incorporate detection of neural error potentials as additional hybrid inputs for further safeguard against unnecessary actions.

The HARMONIE system successfully integrated supervisory control signals into an intelligent robotic framework. Multiple distinct objects were reliably and accurately recognized by the computer vision system, which allowed them to be acted upon by the HARMONIE system. Additionally, the modular nature of the HARMONIE system allowed improvements in the route-planning module between study subjects, with a reduction in completion time from 22.3 to 12.2 s. These improvements allowed the MPL to complete the task more quickly than the average 15 s found in a similar rehabilitative task performed by a BMI using low-level control (Collinger et al. 2013). Unlike directly controlled movement, HARMONIE trials were completed with little temporal variation (<0.1 s), demonstrating the reliability of semi-autonomous movements. The self-feeding task demonstrated how multiple DoF of control can be leveraged to expand the repertoire of actions within a supervisory control framework.

These results further demonstrate the accuracy and stability of ECoG recordings. Without any behavioral training, reach and grasp were robustly and independently decoded and used to control the MPL over the course of several hours of experimental sessions, without significant deterioration in decoding performance. In some cases, there were several days between BMI model training and testing, and the ECoG signals showed remarkable stability over this time scale.

Outlook

The modular nature of the HARMONIE system allows for improvements of its various components to have a significant impact on overall system performance. Just as improved route planning algorithms allowed the present system to complete the entire task in nearly half the time, so will future improvements of several components allow for patients to accomplish even more.

The current system incorporated a single Kinect into the computer vision system; future versions will include the updated Kinect 2 and multiple Kinects to increase the three-dimensional object recognition abilities of the computer vision module. In the future, these sensors could be miniaturized and incorporated into a head-mounted system. Ongoing development in computer vision algorithms will also allow a broader range of objects to be differentiated by the system, which can allow the system to tailor the movement type to the object's characteristics. The current version also used a monitor-based eye tracking system to relay information to and from the patient. Future versions will incorporate glassware-mounted eye-tracking systems with augmented reality displays to allow the patient to interact directly with the environment around them.

Further study into the neuroscience of goal-oriented, cognitive prosthetics will be needed to allow the system to perform a wider variety of ADLs. New tasks can be rapidly incorporated into the HARMONIE system repertoire as the neurophysiological correlates of high-level action goals are mapped and decoded.

References

B.Z. Allison, C. Brunner, V. Kaiser, G.R. Müller-Putz, C. Neuper, G. Pfurtscheller, Toward a hybrid brain–computer interface based on imagined movement and visual attention. J. Neural Eng. 7(2), 026007 (2010)

B.Z. Allison, R. Leeb, C. Brunner, G.R. Müller-Putz, G. Bauernfeind, J.W. Kelly, C. Neuper, Toward smarter BCIs: extending BCIs through hybridization and intelligent control. J. Neural Eng. 9(1), 013001 (2012)

R.A. Andersen, J.W. Burdick, S. Musallam, B. Pesaran, J.G. Cham, Cognitive neural prosthetics. Trends Cogn. Sci. 8(11), 486–493 (2004)

Y. Benjamini, Y. Hochberg, Controlling the false discovery rate: a practical and powerful approach to multiple testing. J. R. Stat. Soc. Ser. B Methodol. 57(1), 289–300 (1995)

J.L. Collinger, B. Wodlinger, J.E. Downey, W. Wang, E.C. Tyler-Kabara, D.J. Weber, A.J.C. McMorland, M. Velliste, M.L. Boninger, A.B. Schwartz, High-performance neuroprosthetic control by an individual with tetraplegia. Lancet 381(9866), 557–564 (2013)

N.E. Crone, D.L. Miglioretti, B. Gordon, R.P. Lesser, Functional mapping of human sensorimotor cortex with electrocorticographic spectral analysis. II. Event-related synchronization in the gamma band. Brain 121(Pt 12), 2301–2315 (1998)

N.E. Crone, D. Boatman, B. Gordon, L. Hao, Induced electrocorticographic gamma activity during auditory perception. Brazier Award-winning article, 2001. Clin. Neurophysiol. 112(4), 565–582 (2001a)

N.E. Crone, L. Hao, J. Hart Jr, D. Boatman, R.P. Lesser, R. Irizarry, B. Gordon, Electrocorticographic gamma activity during word production in spoken and sign language. Neurology 57(11), 2045–2053 (2001b)

N.E. Crone, A. Korzeniewska, P.J. Franaszczuk, Cortical gamma responses: searching high and low. Int. J. Psychophysiol. 79(1), 9–15 (2011)

C. Escolano, J.M. Antelis, J. Minguez, A telepresence mobile robot controlled with a noninvasive brain-computer interface. IEEE Trans. Syst. Man Cybern. Part B Cybern. 42(3), 793–804 (2012)

M.S. Fifer, G. Hotson, B. Wester, D.P. McMullen, Y. Wang, M.S. Johannes, K.D. Katyal, J.B. Helder, M.P. Para, R.J. Vogelstein, W.S. Anderson, N.V. Thakor, N.E. Crone, Simultaneous neural control of simple reaching and grasping with the modular prosthetic limb using intracranial EEG. IEEE Trans. Neural Syst. Rehabil. Eng. 22(3), 695–705 (2014)

F.O. Flemisch, C.A. Adams, S.R. Conway, K.H. Goodrich, M.T. Palmer, P.C. Schutte, The H-metaphor as a guideline for vehicle automation and interaction. NASA Technical Report NASA/TM—2003-212672 (2003)

A. Frisoli, C. Loconsole, D. Leonardis, F. Banno, M. Barsotti, C. Chisari, M. Bergamasco, A new gaze-BCI-driven control of an upper limb exoskeleton for rehabilitation in real-world tasks. IEEE Trans. Syst. Man Cybern. Part C Appl. Rev. **42**(6), 1169–1179 (2012)

K.H. Goodrich, P.C. Schutte, F.O. Flemisch, R.A. Williams, Application of the H-mode, a design and interaction concept for highly automated vehicles, to aircraft, in *25th Digital Avionics Systems Conference, 2006 IEEE/AIAA*, 2006, pp. 1–13

S. Hayati, S.T. Venkataraman, Design and implementation of a robot control system with traded and shared control capability, in *Proceedings IEEE International Conference on Robotics and Automation*, 1989, vol. 3, pp. 1310–1315

I. Iturrate, J.M. Antelis, A. Kubler, J. Minguez, A noninvasive brain-actuated wheelchair based on a P300 neurophysiological protocol and automated navigation. IEEE Trans. Robot. **25**(3), 614–627 (2009)

K.D. Katyal, M.S. Johannes, T.G. McGee, A.J. Harris, R.S. Armiger, A.H. Firpi, D. McMullen, G. Hotson, M.S. Fifer, N.E. Crone, R.J. Vogelstein, B.A. Wester, HARMONIE: A multimodal control framework for human assistive robotics, in *2013 6th International IEEE/EMBS Conference on Neural Engineering (NER)*, 2013, pp. 1274–1278

B.H. Kim, M. Kim, S. Jo, Quadcopter flight control using a low-cost hybrid interface with EEG-based classification and eye tracking. Comput. Biol. Med. **51**, 82–92 (2014)

J.P. Lachaux, N. George, C. Tallon-Baudry, J. Martinerie, L. Hugueville, L. Minotti, P. Kahane, B. Renault, The many faces of the gamma band response to complex visual stimuli. Neuroimage **25**(2), 491–501 (2005)

E.C. Lee, J.C. Woo, J.H. Kim, M. Whang, K.R. Park, A brain–computer interface method combined with eye tracking for 3D interaction. J. Neurosci. Methods **190**(2), 289–298 (2010)

E.C. Leuthardt, G. Schalk, J.R. Wolpaw, J.G. Ojemann, D.W. Moran, A brain-computer interface using electrocorticographic signals in humans. J. Neural Eng. **1**(2), 63–71 (2004)

D.P. McMullen, G. Hotson, K.D. Katyal, B.A. Wester, M.S. Fifer, T.G. McGee, A. Harris, M.S. Johannes, R.J. Vogelstein, A.D. Ravitz, W.S. Anderson, N.V. Thakor, N.E. Crone, Demonstration of a semi-autonomous hybrid brain-machine interface using human intracranial eeg, eye tracking, and computer vision to control a robotic upper limb prosthetic. IEEE Trans. Neural Syst. Rehabil. Eng. **22**(4), 784–796 (2014)

T. Milekovic, T. Ball, A. Schulze-Bonhage, A. Aertsen, C. Mehring, Detection of error related neuronal responses recorded by electrocorticography in humans during continuous movements. PLoS One **8**(2), e55235 (2013)

J.D.R. Millán, R. Rupp, G.R. Müller-Putz, R. Murray-Smith, C. Giugliemma, M. Tangermann, C. Vidaurre, F. Cincotti, A. Kübler, R. Leeb, C. Neuper, K.-R. Müller, D. Mattia, Combining brain–computer interfaces and assistive technologies: state-of-the-art and challenges. Front. Neuroprosthetics **4**, 161 (2010)

S. Musallam, B.D. Corneil, B. Greger, H. Scherberger, R.A. Andersen, cognitive control signals for neural prosthetics. Science **305**(5681), 258–262 (2004)

G. Pfurtscheller, B.Z. Allison, C. Brunner, G. Bauernfeind, T. Solis-Escalante, R. Scherer, T.O. Zander, G. Mueller-Putz, C. Neuper, N. Birbaumer, The hybrid BCI. Front. Neurosci. **4**, 30 (2010a)

G. Pfurtscheller, T. Solis-Escalante, R. Ortner, P. Linortner, G.R. Muller-Putz, Self-paced operation of an SSVEP-based orthosis with and without an imagery-based 'brain switch:' a feasibility study towards a hybrid BCI. IEEE Trans. Neural Syst. Rehabil. Eng. **18**(4), 409–414 (2010b)

S. Ray, E. Niebur, S.S. Hsiao, A. Sinai, N.E. Crone, High-frequency gamma activity (80-150 Hz) is increased in human cortex during selective attention. Clin. Neurophysiol. **119**(1), 116–133 (2008)

A.S. Royer, B. He, Goal selection versus process control in a brain–computer interface based on sensorimotor rhythms. J. Neural Eng. **6**(1), 016005 (2009)

A.S. Royer, M.L. Rose, B. He, Goal selection versus process control while learning to use a brain–computer interface. J. Neural Eng. **8**(3), 036012 (2011)

G. Schalk, K.J. Miller, N.R. Anderson, J.A. Wilson, M.D. Smyth, J.G. Ojemann, D.W. Moran, J. R. Wolpaw, E.C. Leuthardt, Two-dimensional movement control using electrocorticographic signals in humans. J. Neural Eng. **5**(1), 75–84 (2008)

R. Vinjamuri, D.J. Weber, Z.-H. Mao, J.L. Collinger, A.D. Degenhart, J.W. Kelly, M.L. Boninger, E.C. Tyler-Kabara, W. Wang, Toward synergy-based brain-machine interfaces. IEEE Trans. Inf. Technol. Biomed. Publ. IEEE Eng. Med. Biol. Soc. **15**(5), 726–736 (2011)

W. Wang, J.L. Collinger, A.D. Degenhart, E.C. Tyler-Kabara, A.B. Schwartz, D.W. Moran, D. J. Weber, B. Wodlinger, R.K. Vinjamuri, R.C. Ashmore, J.W. Kelly, M.L. Boninger, An electrocorticographic brain interface in an individual with tetraplegia. PLoS One **8**(2), e55344 (2013)

J.R. Wolpaw, Brain–computer interfaces as new brain output pathways. J. Physiol. **579**(3), 613–619 (2007)

T. Yanagisawa, M. Hirata, Y. Saitoh, H. Kishima, K. Matsushita, T. Goto, R. Fukuma, H. Yokoi, Y. Kamitani, T. Yoshimine, Electrocorticographic control of a prosthetic arm in paralyzed patients. Ann. Neurol. **71**(3), 353–361 (2012)

T.O. Zander, M. Gaertner, C. Kothe, R. Vilimek, Combining eye gaze input with a brain-computer interface for touchless human-computer interaction. Int. J. Hum. Comput. Interact. **27**(1), 38–51 (2010)

Near-Instantaneous Classification of Perceptual States from Cortical Surface Recordings

Kai J. Miller, Gerwin Schalk, Dora Hermes, Jeffrey G. Ojemann and Rajesh P.N. Rao

Abstract Human visual processing is of such complexity that, despite decades of focused research, many basic questions remain unanswered. Although we know that the inferotemporal cortex is a key region in object recognition, we don't fully understand its physiologic role in brain function, nor do we have the full set of tools to explore this question. Here we show that electrical potentials from the surface of the human brain contain enough information to decode a subject's perceptual state accurately, and with fine temporal precision. Electrocorticographic (ECoG) arrays were placed over the inferotemporal cortical areas of seven subjects. Pictures of faces and houses were quickly presented while each subject performed a simple visual task. Results showed that two well-known types of brain signals—event-averaged broadband power and event-averaged raw potential—can independently or together be used to classify the presented image. When applied to continuously recorded brain activity, our decoding technique could accurately

K.J. Miller (✉)
Department of Neurosurgery, Stanford University, Stanford, CA, USA
e-mail: kai.miller@stanford.edu

K.J. Miller · J.G. Ojemann · R.P.N. Rao (✉)
Program in Neurobiology and Behavior, University of Washington, Seattle, WA, USA
e-mail: rao@cs.washington.edu

G. Schalk
National Center for Adaptive Neurotechnologies, Wadsworth Center, Albany, NY, USA

D. Hermes
Department of Psychology, Stanford University, Stanford, CA, USA

J.G. Ojemann
Department of Neurological Surgery, University of Washington, Seattle, WA, USA

R.P.N. Rao
Computer Science and Engineering, University of Washington, Seattle, WA, USA

J.G. Ojemann · R.P.N. Rao
Center for Sensorimotor Neural Engineering, University of Washington, Seattle, WA, USA

© The Author(s) 2015
C. Guger et al. (eds.), *Brain-Computer Interface Research*,
SpringerBriefs in Electrical and Computer Engineering,
DOI 10.1007/978-3-319-25190-5_10

predict whether each stimulus was a face, house, or neither, with ∼20 ms timing error. These results provide a roadmap for improved brain-computer interfacing tools to help neurosurgeons, research scientists, engineers, and, ultimately, patients.

Keywords Human vision · Electrocorticography · Broadband power · Event-related potential · Fusiform cortex

Introduction

Throughout each day, people casually see and recognize countless faces, objects, buildings, and other images. These images are represented two-dimensionally on the retina, and the brain has several processing stages that transform the low-level content of each image into the concept of someone or something that we recognize. One of the critical brain regions for advanced stages of object recognition is the inferotemporal cortex. As noted in the introductory chapter, both noninvasive and invasive methods may be used to study brain activity, and both methods have been used to study object recognition in the inferotemoral cortex (Guger et al. in press).

Many groups have used noninvasive methods such as electroencephalography (EEG—with high temporal fidelity, but poor spatial localization) and functional magnetic resonance imaging (fMRI—with very high spatial localization of function, but poor temporal fidelity) to identify areas of the inferotemporal cortex and nearby visual areas that respond to faces, words, and other objects (Puce et al. 1995; Kanwisher et al. 1997; Aguirre et al. 1998; Epstein and Kanwisher 1998). Electrocorticography (ECoG) has emerged as a method to study brain surface activity with high temporal fidelity as well as high spatial resolution, but at the cost of incomplete cortical coverage (Miller et al. 2009, 2014, 2015a).

More recent research has used ECoG, often in combination with other methods, to study category-specific responses in the inferotemporal cortex (Jacques et al. 2015; Miller et al. 2015a, b; Engell and McCarthy 2011, 2014; Engell et al. 2012). These studies have provided additional details about the exact regions, nature, and timing of object recognition, due largely to their ability to assess population-level activity by measuring broadband power, which is not generally measurable with noninvasive methods. Researchers have even used ECoG and other electrical signals to classify presented stimuli into different categories (Simanova et al. 2010, van de Nieuwenhuijzen et al. 2013). However, these studies assumed that the stimulus onset time (the moment that a visual stimulus appeared) was known. Although this may be useful in lab or hospital settings, it is of limited utility in natural environments.

In our study, seven participants viewed two types of images (faces or houses) while we recorded ECoG activity from different areas of the inferotemporal cortex. We explored whether two different types of signals derived from ECoG data (broadband power and event-related potentials or ERPs) would each provide different types of information about object perception. In order to examine this, we developed a template-projection method to discriminate face versus house stimuli using both types of signals, and applied it to the data when the specific time of visual stimulation was either defined or undefined.

Methods

Subjects

This study is comprehensively described in an emerging concurrent manuscript (Miller et al. 2015b). We collected data from seven patients at Harborview Hospital in Seattle, WA. These patients all had medically-refractory epilepsy. In the course of their care, electrodes were placed over different areas of the cortex to help them localize their seizures and monitor them for extended periods. Each subject gave informed consent to participate through an experimental protocol that was approved by an Institutional Review Board (IRB). All patient data were anonymized according to HIPAA guidelines.

Recording

We collected data using subdural platinum electrode arrays (Ad-Tech, Racine, WI), processed through Synamps2 amplifiers (Neuroscan, El Paso, TX) in parallel with clinical recording. Stimuli were presented on a monitor approximately 1 m from each patient's bedside. Structural pre-operative MRIs were co-registered to post-operative CTs to better identify the electrodes' locations relative to major brain structures (Hermes et al. 2010). The BCI2000 software program (Schalk et al. 2004; Schalk and Mellinger 2010) was responsible for presenting stimuli, recording data in tandem with Neuroscan software, and real-time processing. Data were sampled at 1000 Hz, with a reference and ground over the scalp, and filtered from 0.15–200 Hz.

Task

We presented subjects with 10-cm wide grayscale images of faces or houses that were matched for luminance and contrast. Each stimulus remained on the monitor for 400 ms, and there was a blank-screen delay of 400 ms before the next image appeared. Each patient participated in three experimental runs, and each run had

100 pictures (50 of each type—faces or houses, with the order chosen randomly). To ensure that subjects paid attention, we presented an upside-down house once during each run, and asked subjects to report when it appeared. Subjects very rarely made any errors in this task (~ 3–5 across all subjects), which confirms that they were paying attention to the stimulus discrimination task.

Signal Processing and Classification

We employed several different stages of signal processing and classification after data collection. After removing electrodes with epileptic or artifactual activity, ECoG signals were re-referenced to the average signal across those remaining electrodes [common average reference (CAR)]. We also filtered out signal from 58–62 Hz, to remove ambient 60 Hz environmental noise from electrical potential measurement. Next, we calculated the power spectra of 1-s epochs of the data to estimate the signal's amplitude and phase at each time point. Finally, we separated two kinds of brain activity: rhythmic activity and broadband change (Miller et al. 2014; see also Porat 1997). Rhythmic activity in low frequencies is generally assigned a modulatory role in cortical processing. In contrast, changes in broadband activity are commonly understood to represent the activity of local populations of neurons, which is the center focus of our investigations here. Thus, while brain rhythms are likely of interest in other research, they were isolated and removed here.

To define the average responses of the brain to faces or houses, we created averaged templates of stimulus-related responses, separately for event-related potential (ERP) and event-related broadband (ERBB) signals (see Fig. 1b). We subsequently convolved these templates with the ERP and ERBB signals. The resulting time courses typically peaked at the time of stimulus presentation, and were specific to the type of stimulus (face or house).

We then used these time courses to provide input features for a linear discriminant analysis (LDA, Bishop 1995) that was trained to differentiate faces from houses. This training and subsequent evaluation made use of threefold cross-validation. I.e., we used two thirds of the data to train the classifier, and one third of the data to test the classifier. With this procedure, our classifier examined data from the testing set and tried to determine whether it resulted from a face stimulus or house stimulus. We conducted this classification with data only at the times at visual stimulation, and also with continuous data streams. We did this to assess how well we could identify different brain processes when we know when the stimulus occurs (such as in labs or hospitals) and when we don't (such as in natural environments).

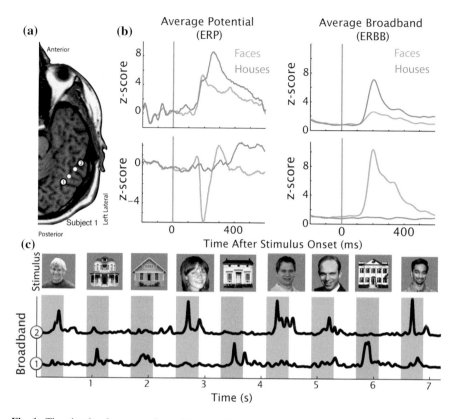

Fig. 1 The visual task presented to subjects (with faces and houses), and the responses elicited at two different electrodes. **a** Locations of four adjacent recording sites for subject 1. Each of the four dots represents one ECoG electrode. The four electrodes are located on a strip that was placed over the ventral temporal area of the brain. Panels **b** and **c** present data from the electrodes labeled 1 and 4. **b** These four graphs show the activity elicited by faces (*green*) and houses (*blue*) over electrodes 1 (*top panel*) and 4 (*bottom panel*). The *two left panels* show the averaged ERP activity over these two sites, while the *two right panels* show the averaged ERBB activity. **c** The *top half* of this panel presents the stimuli presented to all subjects: grayscale faces and houses. Each of these images was present for 400 ms, and the screen was blank (*grey*) for 400 ms before the next image. The subjects were asked to detect an upside-down face or house amidst the stream of upright images. Below these images, we present the broadband spectral change over locations 1 and 4 (from panel **a**) when these images were presented. The *vertical green areas* show times of face image presentation, and the *blue vertical areas* show times of house image presentation. Electrodes 1 and 4 only show strong changes for house and face stimuli, respectively

Results

The following results present classifier accuracy based on different types of information (ERP alone, ERBB alone, or both in combination), and when the stimulus onset time is both known and unknown.

Event-Related Broadband (ERBB) and Event-Related Potential (ERP) Activity: Changes During a Visual Task

Figure 1 shows results from one subject. We chose two electrodes (highlighted in panel a) that show distinct activity in response to face versus house stimuli. Panels b and c both show that electrode 1 exhibits ERP and broadband changes to images of houses, but little response to face images. Electrode 4 shows the opposite pattern.

Figure 1 shows that the ECoG activity in electrodes 1 and 4 (ERPs and broadband power) can be used to differentiate the responses to houses and faces over two different brain regions. To the naked eye, these qualitative differences become obvious about 200 ms after stimulus onset. The following two sections present quantitative assessments (measured by classifier performance) when the classifier is informed about the stimulus onset time (that is, when the image appeared), and when it does not.

Classifier Performance: Known Stimulus Onset Time

First, we explored how well our automated classifier could identify the stimulus class (face or house) with a known stimulus onset time. As described above, we created templates for the ERP and ERBB changes each time a stimulus appeared, with 2/3 of the data used to train the classifier and the remaining 1/3 used to test it. Table 1 presents results across seven subjects.

Table 1 suggests that ERP activity is slightly less useful than broadband power for this classification task when stimulus onset time is known. With only one type of signal, the ERBB-based classifier outperforms the ERP-based classifier with data from five of the seven subjects. Furthermore, the addition of ERP activity only improved the ERBB-based classifier's performance in three of seven subjects, and actually impaired classifier performance in two subjects. However, the classifier attained at least 90 % accuracy in all conditions for all subjects shown here, with only two exceptions. When both templates were combined, the classifier attained over 90 % accuracy with all subjects.

These results demonstrate that ERP and ERBB signals detected in electrocorticography can reliably identify the type of visual stimulus (face or house) in single trials when the timing of the visual stimulus is known. The next section extends these results to the case in which the classifier is not informed about the timing of the stimulus.

Table 1 This table shows classification accuracy when the classifier is aware of the stimulus onset time

Proportion correct	Subject 1	Subject 2	Subject 3	Subject 4	Subject 5	Subject 6	Subject 7
Voltage (ERP)	0.9	0.91	0.77	0.95	0.96	0.96	0.84
Broadband (ERBB)	0.97	0.99	0.98	1	0.96	0.94	0.96
Both together	0.95	1	0.91	1	0.98	0.97	0.96

Results show accuracy when the classifier used ERP, broadband power, or both of these templates in combination

Decoding Stimulus Class and Onset from a Continuous Cortical Data Stream

The computer system in this study recorded each moment when a new stimulus appeared. To assess performance when stimulus onset time is unknown, we tested the classifier with a continuous data stream.

To do this, we applied the classifier to each point in time, which resulted in a time course of posterior probabilities. We then smoothed this time course with a Gaussian filter (sigma = 80 ms). From this smoothed time course, we assigned of predicted stimulus onset time by identifying those local peaks in the posterior probability time course that were spaced at least 320 ms. We considered predicted

Table 2 This table presents classifier accuracy from a continuous data stream, when the stimulus onset is unknown

	Subject 1	Subject 2	Subject 3	Subject 4	Subject 5	Subject 6	Subject 7
Proportion of stimuli captured							
Voltage (ERP)	0.89	0.93	0.8	0.97	0.98	0.95	0.89
Broadband	0.89	0.97	0.92	0.92	0.94	0.89	0.94
Combination	0.94	0.97	0.95	0.98	0.97	0.97	0.97
Temporal accuracy (ms)							
Voltage (ERP)	19 (SD 14)	25 (SD 26)	20 (SD 16)	18 (SD 15)	18 (SD 12)	20 (SD 15)	22 (SD 22)
Broadband	32 (SD 33)	35 (SD 30)	30 (SD 26)	38 (SD 35)	25 (SD 19)	27 (SD 20)	34 (SD 34)
Combination	18 (SD 13)	19 (SD 18)	19 (SD 15)	19 (SD 17)	18 (SD 12)	21 (SD 17)	21 (SD 19)
Proportion of guesses that are incorrect							
Voltage (ERP)	0.11	0.11	0.23	0.06	0.02	0.06	0.16
Broadband	0.17	0.03	0.11	0.11	0.06	0.12	0.06
Combination	0.05	0.04	0.07	0.03	0.03	0.03	0.03

Like Table 1, results are presented across seven subjects based on three analyses: ERP only, broadband power only, and both template types in combination. The top section presents the proportion of stimuli captured by the classifier. The second section presents the mean (and standard deviation in parentheses) of the temporal accuracy—that is, the error in predicting when each stimulus appeared. The third section shows the proportion of guesses that were incorrect (i.e., predicted stimuli at the wrong time, or as the wrong class)

stimulus times as correct if they indeed occurred within 160 ms of the actual stimulus time. Table 2 presents results from this more difficult classification challenge.

The classifier performed only slightly worse when the stimulus onset time was unknown. When using a combination of features, accuracy was still at least 94 % for all subjects, with only 4 % incorrect guesses on average, and the classifier could predict the stimulus onset within about 20 ms.

Unlike the classifier results in Table 1, the ERP activity in Table 2 provides a very important supplement to the ERBB activity. In all three sections of Table 2, the combination of template types performs better than ERBB alone in most or all subjects. ERP activity was especially helpful in predicting temporal accuracy. Thus, these two features of cortical activity contain complementary information about the neural processes underlying visual perception.

Discussion

People can perceive stimuli in one of two environments. In a lab or hospital environment, researchers know exactly when stimuli appear. This can make analysis and classification much easier, but it not realistic in most real-world settings. Previous work in which the stimulus timing was known could predict the stimulus class (in excess of chance) within 100 ms after it appears (Hung et al. 2005; Kiani et al. 2007; Liu et al. 2009; Simanova et al. 2010; van de Nieuwenhuijzen et al. 2013). People can also make eye movements (called saccades) towards an image with a face within 140 ms after the image appears (Kirchner and Thorpe 2006; Crouzet et al. 2010). Our work shows that an automated classifier can attain 96 % accuracy in distinguishing faces versus houses when the stimulus onset is known. This is a substantial improvement over earlier work in this area, where onset timing was pre-defined, with peak accuracies of 72 % with MEG/fMRI (Cichy et al. 2014), 89 % with EEG (Simanova et al. 2010), and 94 % with MEG (van de Nieuwenhuijzen et al. 2013).

However, people usually perceive stimuli in a natural environment, when nobody knows exactly when new objects appear. Stimuli instead arrive continuously, and thus the brain must constantly update its perceptual state. Thus, we studied classification accuracy when the stimulus onset was unknown. Results showed that ECoG data can be used to predict which stimulus occurred (face vs. house vs. blank image) and when it appeared (within about 20 ms accuracy).

We also studied the relative contributions of broadband power changes versus ERP activity versus both signals in combination. To date, these signals have not yet been explored in combination for this purpose. When stimulus onset was known, broadband activity was most effective at predicting which stimulus occurred. However, when we were classifying data in a natural setting, the combination of both signal types led to the highest classification accuracy. This result suggests that these two types of signals each reflect different, but complementary, brain processes involved in object recognition. Specifically, the ERP information seems to convey useful information about the stimulus timing that is not available from broadband activity. Since ERP and broadband signals each reflect different brain processes, this new information could help future researchers learn more about the brain mechanisms responsible for object perception.

Acknowledgments We are grateful to the patients and staff at Harborview Hospital in Seattle. We are grateful for helpful discussions with Kalanit Grill-Spector and Bharathi Jagadeesh. This work was supported by National Aeronautics and Space Administration Graduate Student Research Program (KJM), the NIH (R01-NS065186 (KJM, JGO, RPNR), T32-EY20485 (DH), R01-EB006356 (GS), R01-EB00856 (GS) and P41-EB018783 (GS)), the NSF (EEC-1028725 (RPNR)), the US Army Research Office (W911NF-07-1-0415 (GS), W911NF-08-1-0216 (GS) and W911NF-14-1-0440 (GS)), and Fondazione Neurone (GS).

References

G.K. Aguirre, E. Zarahn, M. D'Esposito, An area within human ventral cortex sensitive to "building" stimuli: evidence and implications. Neuron **21**, 373–383 (1998)

C.M. Bishop, *Neural Networks for Pattern Recognition* (Oxford University Press, Oxford, 1995)

R.M. Cichy, D. Pantazis, A. Oliva, Resolving human object recognition in space and time. Nat. Neurosci. **17**, 455–462 (2014)

S.M. Crouzet, H. Kirchner, S.J. Thorpe, Fast saccades toward faces: face detection in just 100 ms. J. Vis. **10**(16), 11–17 (2010)

A.D. Engell, G. McCarthy, The relationship of gamma oscillations and face-specific ERPs recorded subdurally from occipitotemporal cortex. Cereb. Cortex **21**, 1213–1221 (2011)

A.D. Engell, G. McCarthy, Repetition suppression of face-selective evoked and induced EEG recorded from human cortex. Human Brain Map (2014). doi:10.1002/hbm.22467

A.D. Engell, S. Huettel, G. McCarthy, The fMRI BOLD signal tracks electrophysiological spectral perturbations, not event-related potentials. NeuroImage **59**, 2600–2606 (2012)

R. Epstein, N. Kanwisher, A cortical representation of the local visual environment. Nature **392**, 598–601 (1998)

C. Guger, B.Z. Allison, G.R. Müller-Putz, in *Brain-Computer Interface Research: A State-of-the-Art Summary 4* (in press)

D. Hermes, K.J. Miller, H.J. Noordmans, M.J. Vansteensel, N.F. Ramsey, Automated electrocorticographic electrode localization on individually rendered brain surfaces. J. Neurosci. Methods **185**, 293–298 (2010)

C.P. Hung, G. Kreiman, T. Poggio, J.J. DiCarlo, Fast readout of object identity from macaque inferior temporal cortex. Science **310**, 863–866 (2005)

C. Jacques, N. Witthoft, K.S. Weiner, B.L. Foster, V. Rangarajan, D. Hermes, K.J. Miller, J. Parvizi, K. Grill-Spector, Corresponding ECoG and fMRI category-selective signals in Human ventral temporal cortex. Neuropsychologia (2015)

N. Kanwisher, J. McDermott, M.M. Chun, The fusiform face area: a module in human extrastriate cortex specialized for face perception. J. Neurosci. **17**, 4302–4311 (1997)

R. Kiani, H. Esteky, K. Mirpour, K. Tanaka, Object category structure in response patterns of neuronal population in monkey inferior temporal cortex. J. Neurophysiol. **97**, 4296–4309 (2007)

H. Kirchner, S.J. Thorpe, Ultra-rapid object detection with saccadic eye movements: visual processing speed revisited. Vis. Res. **46**, 1762–1776 (2006)

H. Liu, Y. Agam, J.R. Madsen, G. Kreiman, Timing, timing, timing: fast decoding of object information from intracranial field potentials in human visual cortex. Neuron **62**, 281–290 (2009)

K.J. Miller, L.B. Sorensen, J.G. Ojemann, M. den Nijs, Power-law scaling in the brain surface electric potential. PLoS Comput. Biol. **5**, e1000609 (2009)

K.J. Miller, C.J. Honey, D. Hermes, R.P. Rao, M. denNijs, J.G. Ojemann, Broadband changes in the cortical surface potential track activation of functionally diverse neuronal populations. NeuroImage **85**(Pt 2), 711–720 (2014)

K.J. Miller, D. Hermes, N. Witthoft, R.P. Rao, J.G. Ojemann, The physiology of perception in human temporal lobe is specialized for contextual novelty. J. Neurophysiol. **114**, 256–263 (2015a)

K.J. Miller, G. Schalk, D. Hermes, J.G. Ojemann, R.P.N. Rao, *Spontaneous Decoding of the Timing and Content of Human Object Perception from Cortical Surface Recordings Reveals Complementary Information in the Event-Related Potential and Broadband Spectral Change* (2015b)

B. Porat, *A Course in Digital Signal Processing* (Wiley, New York, 1997)

A. Puce, T. Allison, J.C. Gore, G. McCarthy, Face-sensitive regions in human extrastriate cortex studied by functional MRI. J. Neurophysiol. **74**, 1192–1199 (1995)

G. Schalk, J. Mellinger, *A Practical Guide to Brain-Computer Interfacing with BCI2000* (Springer, London, 2010)

G. Schalk, D.J. McFarland, T. Hinterberger, N. Birbaumer, J.R. Wolpaw, BCI2000: a general-purpose brain-computer interface (BCI) system. IEEE Trans. Biomed. Eng. **51**(6), 1034–1043 (2004)

I. Simanova, M. van Gerven, R. Oostenveld, P. Hagoort, Identifying object categories from event-related EEG: toward decoding of conceptual representations. PLoS ONE **5**, e14465 (2010)

M.E. van de Nieuwenhuijzen, A.R. Backus, A. Bahramisharif, C.F. Doeller, O. Jensen, M.A. van Gerven, MEG-based decoding of the spatiotemporal dynamics of visual category perception. NeuroImage **83**, 1063–1073 (2013)

The Changing Brain: Bidirectional Learning Between Algorithm and User

N. Mrachacz-Kersting, N. Jiang, S. Aliakbaryhosseinabadi, R. Xu,
L. Petrini, R. Lontis, K. Dremstrup and D. Farina

Abstract In 2013–2014 we have advanced our MRCP-based BCI by demonstrating:
(1) the ability to detect movement intent during dynamic tasks; (2) better detection
accuracy than conventional approaches by implementing the locality preserving
projection (LPP) approach; (3) the ability to use a single channel for accurate detection; and (4) enhanced neuroplasticity by driving a robotic device in an online mode.
To realize our final goal of an at home system, we have characterized alterations
during single session use in our extracted signal when the user is undergoing complex
learning or experiencing significant attentional shifts—all seriously affecting the
detection of user intent. Learning enhances the variability of MRCP at specific
recording sites while attention shifts result in a more global increase in signal
variability. With the results presented, we are working towards an adaptive brain–
computer interface where bidirectional learning (either user or algorithm) is possible.

Keywords Brain computer interface · Motor imagery · Movement related cortical
potentials

Introduction

Brain Computer Interfaces (BCIs) intended for restoration of lost motor function are
designed to induce neuroplasticity of those areas damaged by a central nervous
system lesion such as stroke (Daly and Wolpaw 2008; Shih et al. 2012). Over the

N. Mrachacz-Kersting (✉) · S. Aliakbaryhosseinabadi · L. Petrini · R. Lontis · K. Dremstrup
Center for Sensory-Motor Interaction, Department of Health Science and Technology,
Aalborg University, 9220 Aalborg, Denmark
e-mail: nm@hst.aau.dk

N. Jiang · R. Xu · D. Farina
Neurorehabilitation Engineering Bernstein, Center for Computational Neuroscience
University Medical Center, Göttingen, Germany

© The Author(s) 2015
C. Guger et al. (eds.), *Brain-Computer Interface Research*,
SpringerBriefs in Electrical and Computer Engineering,
DOI 10.1007/978-3-319-25190-5_11

past seven years, several research groups, including our own, have focused their efforts toward the design of a robust algorithm that can detect user intent and transfer this information for device control that reproduces the intended movement. By continuous pairing of user intent and artificial reproduction of that movement, such BCIs follow the principle of Hebbian association (Hebb 1949) that underlies motor learning leading to neuroplasticity and associated functional changes (Ramos-Murguialday et al. 2013; Daly et al. 2009; Li et al. 2014). In principle, a number of signals may be extracted from continuous non-invasive electroen-cephalographic (EEG) recordings such as event related Sensory-motor rhythms (SMR) or slow movement related cortical potentials (MRCPs). Here, the importance of early detection of movement intent has been underlined in several manuscripts where MRCP was shown to outperform SMR in terms of accurately timed device control (Ramos-Murguialday et al. 2013; Niazi et al. 2011, 2012; Xu et al. 2014a, b; Hashimoto and Ushiba 2013; Pichiorri et al. 2011), and consequently in rapid induction of neuroplasticity (Niazi et al. 2012; Xu et al. 2014b) that accompanies improvements in functionality (Ramos-Murguialday et al. 2013; Pichiorri et al. 2011).

In 2013–2014, with the ultimate aim of bringing a BCI for neuromodulation to the market, we have been working in several directions including improved detection (enhancing the performance of the algorithm) (Xu et al. 2014a, b), detection during dynamic movements such as gait initiation (Jiang et al. 2014), reducing the number of channels required to detect user intent (altering the recording device) (Jochumsen et al. 2014) and manipulating the artificially pro-duced movement by using either peripheral nerve electric stimulation (ES) to activate the weak muscles directly or implementing a motorized foot ankle orthotic device to passively replicate the movement. The latter provides a more physio-logical input to the brain than ES when an imagined movement occurs. From this body of work, it is reasonable to conclude that: (I) Detection is significantly enhanced by using the locality preserving projection (LPP) approach, outper-forming other linear methods, e.g. LDA and PCA. Xu et al. (2014a) demonstrated LPP's superiority over the classic matched filter in online detection of motor intention detection using MRCP, with higher accuracy and lower latency; (II) Detection accuracy is not significantly affected by a reduction in the number of channels (Jochumsen et al. 2014) or task complexity (Jiang et al. 2014); and (III) The more physiologically valid the artificially produced movement is, the greater the induced neuroplasticity (Xu et al. 2014b).

Following these achievements, our focus has recently shifted to understand more thoroughly how the firing of neuronal assemblies within the human motor cortex and surrounding areas is altered during the learning of a new or indeed the reha-bilitation of an old motor task or with alterations in attention—both imperative for a robust BCI (Leuthardt et al. 2009; Vaadia and Birbaumer 2009).

Alterations in MRCP due to the Induction of Plasticity

Since the seminal work by Pascual-Leone and colleagues (Pascual-Leone et al. 1994), it has generally become accepted that, as complex motor tasks are learned, the area that is activated during task performance expands. In that study, subjects were required to learn a complex five finger sequence on a piano using only one hand over 5 days. The instructions were to maintain the tempo of 60 beats per minute (provided using auditory feedback). Non-invasive transcranial magnetic stimulation was used to elicit motor evoked potentials (MEPs) in the long flexor and extensor muscles of the hand bilaterally. Incrementally across days, the size of the cortical representation of both muscle groups increased significantly only for the practiced hand and only following the completion of practice on that day. If subjects were asked to only mentally practice the same sequence, these alterations in cortical maps were similar. For patients unable to perform voluntary movements, this poses a distinct advantage, since the mere mental imagery (MI) of the movement will result in changes at the cortical level similar to actual movement performance. However, such changes cannot continue indefinitely; that is, the cortical maps will not increase ad infinitum, but decrease again once the motor task learned has manifested (Karni et al. 1998; Pascual-Leone et al. 2005). In recent years, the concept of homeostatic metaplasticity has emerged in an effort to understand the changes observed with motor skill learning. Ziemann et al. (2006) showed that homeostatic plasticity affects motor learning. Here, motor learning was enhanced if subjects performed the practice of rapid thumb flexion, subsequent to a long-term-depression (LTD)-like plasticity inducing protocol. LTD weakens synaptic connections by a continuously unassociated pairing of two inputs to the post-synaptic neuron (Cooke and Bliss 2006; Bliss and Cooke 2011).

For BCI performance, both plasticity and homeostatic metaplasticity have significant implications, since the algorithm used to extract user intent is based on a training set recorded at the start of a BCI session. This has already been recognized by other groups working within BCI (Shih et al. 2012; Leuthardt et al. 2009; Vaadia et al. 2009). It is thus generally acknowledged that, if the BCI user is subjected to plasticity induction, then the algorithm performance may be seriously affected. Indeed, both skill acquisition using the finger flexors or extensors (Demiralp et al. 1990), the index finger (Dirnberger et al. 2004) or explosive strength training involving the hamstring muscle group (Falvo et al. 2010) affect the MRCP amplitude and slope. Explosive strength training can also affect the onset of the MRCP, being 28 % earlier following training (Falvo et al. 2010). It is thus surprising why efforts have failed to concentrate on developing an adaptable algorithm capable of detecting alterations in brain activity due to plasticity. Daly was the first to propose a BCI intended for neuromodulation and thus plasticity induction when she reported on a single case where functional electrical stimulation was triggered using brain signals captured from the lesioned hemisphere (Daly et al. 2009). Following 3 weeks of training, the patient had recovered some volitional isolated index finger extension movement abilities.

Since her work, several groups have applied BCI training for neuromodulation in particular to stroke patients (Silvoni et al. 2011). In all of these studies, plasticity has been quantified by functional improvements (Daly et al. 2009; Li et al. 2014; Ang et al. 2010; Broetz et al. 2010; Cincotti et al. 2012; Mukaino et al. 2014; Young et al. 2014), changes in extracted EEG parameters (Li et al. 2014; Broetz et al. 2010; Cincotti et al. 2012; Yilmaz et al. 2013) or through fMRI (Mukaino et al. 2014; Young et al. 2014; Song et al. 2014). However, none of these studies have tried to adapt the extraction algorithm to the new user state. While the positive changes reported on user performance of the BCI in some of these studies are encouraging, it is possible that performance may have been further enhanced (and at an earlier stage) if changes in brain state had been integrated into the BCI extraction algorithm.

In our BCI paradigm using MRCPs, we have targeted the tibialis anterior (TA) muscle to counteract the effects of foot-drop in chronic stroke patients (Farina 2012). In particular during human walking, the TA is engaged in dorsi-flexing the foot during the swing phase of the gait cycle. During the late swing phase it acts to prepare the foot for heel strike and is primarily controlled via the spinal reciprocal inhibitory pathway (Petersen et al. 1999). Nielsen's group has demonstrated that significant plasticity of the corticospinal tract to the TA, as assessed through alterations in MEP size, may be induced by skilled motor learning but not by either passive or simple movements (Perez et al. 2004). In the first series of experiments, we set out to investigate if the same skilled motor learning (as opposed to a simple motor task) would have an effect on our primary EEG signal of interest, the MRCP.

The validated complex motor learning task leading to neuroplasticity (also used by Nielsen and colleagues (Perez et al. 2004), versus a control condition, was performed by eight subjects over a 32 min period. For both, subjects were seated comfortably in an armchair with the dominant leg flexed in the hip (120°), the knee (160°) and the ankle (110° of plantar flexion). The foot was resting on a foot-plate and a computer screen was positioned approximately 1.5 m in front of the subjects at eye level. Figure 1 shows the experimental set-up.

The learning task was comprised of a series of six randomized figures, each sketching a different series of combinations of dorsi- and plantarflexion movements (Fig. 2a). Upon appearance of the trace on the screen, a countdown of 3 s was visually shown. On the word 'go', the activity of the subject's TA was displayed in real-time overlaying the respective trace. Electromyography (EMG) was used to quantify this muscle activity. Subjects were instructed to follow these traces by controlling the activation level of their TA muscle. Each trace lasted 3–4 s and a single training run lasted for 4 min followed by a 2 min rest period. A total of eight training runs were completed, leading to a total training time of 32 min. The control condition was identical to the learning task, with the exception that subjects did not follow a trace but instead simply had to perform single dorsiflexions for the same period of time. MRCPs were extracted using LPP from the following electrodes: Fz, FC1, FC2, C3, Cz, C4, CP1, CP2, Pz during a complex dorsiflexion task performed pre and post the complex learning or the control condition. Task performance was assessed by quantifying the correlation between the displayed trace and the actual task performance during the pre and post measures.

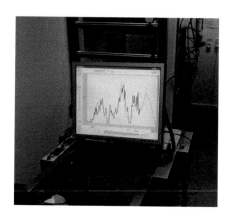

Fig. 1 Experimental set-up

The effect of the training for one subject is illustrated in Fig. 2b–e. Figure 2b, c shows the data for one of the traces selected for analysis, while Fig. 2d, e shows the scatter plot between the exerted TA activation and the target level for each interval of time. For this subject, the correlation between shown traces and actual subject performance increased from 0.84 to 0.91. Only the complex motor task training group significantly improved task performance. The correlation between the trace and the performance changed from 0.84 to 0.91 (p = 0.0005), while that of controls remained the same at 0.81 (p = 0.72).

An MRCP variability curve was calculated from individual pre and post learning trials (see Fig. 3 for a representative sample from C3 and Cz, pre and post mean and variability traces). Complex motor task learning induced increases in variability (20–36 %) across specific channels (C3, FC1, FC2 and Fz), while no such changes were seen in the control condition. As a final step, we investigated the robustness of the detection algorithm. This was expressed as the rate of true positives (TPR) and false positives (FP) as a percentage. For complex learning, these were: TPR: 77.9 ± 7.6 (pre) and 55.0 ± 22.5 (post), FP: 24.1 ± 4.0 (pre) and 44.1 ± 8.6 (post); for the control task they were: TPR: 79.4 ± 10.1 (pre) and 63.3 ± 22.2 (post), FP: 32.6 ± 16.5 (pre) and 45.4 ± 19.0 (post).

Thus, plasticity induction significantly affected detection performance and thus usability of our BCI. What is promising is that not all channels were affected, but only several key channels located in the frontal cortex. This suggests that it may be possible to identify altered brain state due to plasticity versus (for example) changes due to attention, which are known to affect more posterior channels.

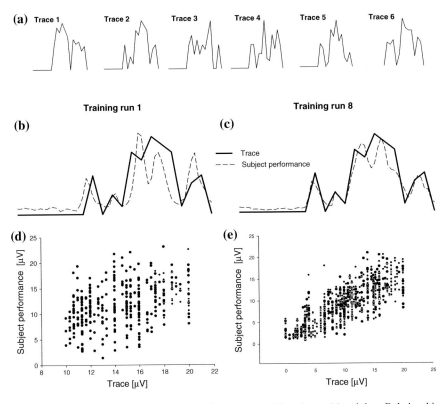

Fig. 2 The 6 traces (**a**) trace and subject performance pre (**b**) and post (**c**) training. Relationship between trace and subject data for all data points pre (**d**) and post (**e**) training (n = 1)

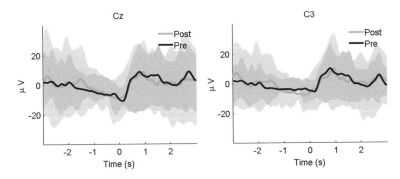

Fig. 3 C3 and Cz pre and post complex learning in n = 1. The *thick lines* are the pre and post MRCP (averaged over 30 trials), and the corresponding *shaded areas* indicate the MRCP variability (±STD) for each case

Alterations in MRCP due to Shifts in Attention

As indicated in the introduction, aside from plasticity, an equally important aspect is the alteration in attention to the task by the user. In a BCI used at home or in a community setting, the sterile conditions of the laboratory cease to exist. Hence, it is likely that distractions from the surrounding environment divert the attention of the BCI user in an unpredictable manner, as has been shown for the P300. The BCI user is likely to be constantly exposed to various types of sensory stimuli arising both from sensors detecting changes within the body (e.g. proprioceptors) and those detecting changes in the environment (e.g. vision). All such signals are attenuated as they progress from the periphery and via the thalamus to the sensory cortex, where they are integrated into movement production. Humans thus selectively process simultaneous sources of information, yet this differs across individuals due to differences in attention. These may be related to age, neurological status, fatigue, stress, practice, and other factors.

In an effort to better understand how attention affects task performance, several models have been proposed (McDowd 2007). Fundamentally, these differ in what modulates attention: the task demands or the performer's goal. For BCI performance, this poses a seminal problem, since if it is the task demands that dictate attention, then the BCI needs to be tuned to the user's attentional capabilities. In the case of the performer's goals dictating attention, the requirement shifts so that both BCI and user need to be tuned to meet the demands of the user's goals. In our BCI paradigm, we were initially interested in whether changes in the user's attention would affect our extracted parameter, the MRCP, and therefore also the performance of our BCI algorithm.

A subject's attention level can be experimentally manipulated using standardized techniques such as the "oddball task" (Tome et al. 2015; Yurgil and Golob 2013). Changes in the attention level can then be monitored using event-related brain potentials (ERP) from EEG and behavioral responses (reaction times and accuracy). One type of oddball task presents three different stimuli in a random sequence, with one (target = T) occurring less frequently than the others (standard = S or deviant + D). The subject is instructed to respond only to the target stimulus. This task is able to elicit a very specific ERP, labeled as P300, which is thought to reflect the allocation of cognitive attention resources to a task. The P300 amplitude is an index of the amount of resources invested to identify a target stimulus, whereas the P300 latency is an index of stimulus evaluation time. When task conditions are not demanding, the P300 amplitude is large and peak latency is short. For tasks that require greater amounts of attentional resources, the P300 amplitude is smaller and peak latency is longer, since processing resources are used for task performance (Polich 2007). In our second study, we used this knowledge to shift the attention of the user by implementing a validated auditory oddball task (5 min duration). This required participants (both healthy and chronic stroke patients) to press a mouse button upon hearing one of three sounds of various frequencies.

The P300 was used to quantify the attentional state. MRCPs were extracted using LPP from the following electrodes: Fz, FC1, FC2, C3, Cz, C4, CP1, CP2, Pz while participants performed a dorsiflexion of the ankle joint in a series of 30 trials both prior to and following the oddball task or during a control condition. A significant decrease in the value of the MRCP peak negativity (PN) occurred at the central channel locations due to the attention shift ($F_{(1,12)} = 13.14$, $p = 0.03$). The time of PN was not significantly altered. The performance of the detection algorithm was quantified as the rate of true positives (TPR) and false positives (FP) as a percentage. The TPR was 81 ± 7 and 64 ± 8.7 % (healthy, control vs. attention shift), 71.7 ± 4.4 and 59 ± 3.1 % (chronic stroke patients, control vs. attention shift) and FP/min: 5 and 8.7 (healthy, control vs. attention shift), 7 and 11 (chronic stroke patients, control vs. attention shift). Thus, regarding the induction of plasticity, attention shifts led to a significant decline in the performance of our BCI. Further studies are currently being conducted to localize precisely the attentional effects on the cortex.

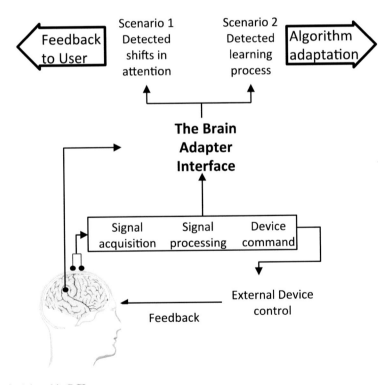

Fig. 4 Adaptable BCI

Long-Term Perspectives

The schematic in Fig. 4 outlines our proposed approach to develop an adaptable algorithm capable of detecting altered states of the user. To realize this, several preliminary studies have been completed that characterized the alterations induced by plasticity and attention shifts on parameters of the MRCP required by our current algorithm based on LPP, where signal variability is of central importance (Xu et al. 2014a).

Plasticity and attention affect MRCP in significantly different ways, yet both cause significant declines in the performance of the non-adaptive algorithm. While plasticity induction leads to an increased variability of the signal at electrodes located frontally, attentional shifts induce a more global change in all underlying electrodes. The consequence is that we are able to differentiate between these two brain states, and are currently working towards an integration in our algorithm.

As such, it is our vision that MRCP in the future will not be limited to neuro-modulation, but will offer a robust signal modality for accurate device control in other BCI applications.

Acknowledgements We wish to acknowledge our subjects and all of the students from the laboratory, both past and present.

References

K.K. Ang, C. Guan, K.S. Chua, B.T. Ang, C. Kuah, C. Wang, K.S. Phua, Z.Y. Chin, H. Zhang, Clinical study of neurorehabilitation in stroke using EEG-based motor imagery brain-computer interface with robotic feedback. Conf. Proc. IEEE Eng. Med. Biol. Soc. **2010**, 5549–5552 (2010)

T.V. Bliss, S.F. Cooke, Long-term potentiation and long-term depression: a clinical perspective. Clinics (Sao Paulo) **66**(1), 3–17 (2011)

D. Broetz, C. Braun, C. Weber, S.R. Soekadar, A. Caria, N. Birbaumer, Combination of brain-computer interface training and goal-directed physical therapy in chronic stroke: a case report. Neurorehabil. Neural Repair. **24**, 674–679 (2010)

F. Cincotti, F. Pichiorri, P. Arico, F. Aloise, F. Leotta, F. de Vico Fallani, R. Millan Jdel, M. Molinari, D. Mattia, EEG-based Brain-Computer Interface to support post-stroke motor rehabilitation of the upper limb. Conf. Proc. IEEE Eng. Med. Biol. Soc. **2012**, 4112–4115 (2012)

S.F. Cooke, T.V.P. Bliss, Plasticity in the human central nervous system. Brain **129**, 1659–1673 (2006)

J.J. Daly, J.R. Wolpaw, Brain-computer interfaces in neurological rehabilitation. Lancet Neurol. **7**, 1032–1043 (2008)

J.J. Daly, R. Cheng, J. Rogers, K. Litinas, K. Hrovat, M. Dohring, Feasibility of a new application of noninvasive Brain Computer Interface (BCI): a case study of training for recovery of volitional motor control after stroke. J. Neurol. Phys. Ther. **33**, 203–211 (2009)

T. Demiralp, S. Karamursel, Y.E. Karakullukcu, N. Gokhan, Movement-related cortical potentials: their relationship to the laterality, complexity and learning of a movement. Int. J. Neurosci. **51**, 153–162 (1990)

G. Dirnberger, C. Duregger, G. Lindinger, W. Lang, Habituation in a simple repetitive motor task: a study with movement-related cortical potentials. Clin. Neurophysiol. **115**, 378–384 (2004)

M.J. Falvo, E.J. Sirevaag, J.W. Rohrbaugh, G.M. Earhart, Resistance training induces supraspinal adaptations: evidence from movement-related cortical potentials. Eur. J. Appl. Physiol. **109**, 923–933 (2010)

D. Farina, A novel brain-computer interface for chronic stroke patients, Toledo, Spain (2012)

Y. Hashimoto, J. Ushiba, EEG-based classification of imaginary left and right foot movements using beta rebound. Clin. Neurophysiol. (2013)

D.O. Hebb, *The organization of behavior: a neuropsychological theory* (Lawrence Erlbaum Associates Inc, Mahwah, 1949)

N. Jiang, L. Gizzi, N. Mrachacz-Kersting, K. Dremstrup, D. Farina, A brain computer interface for single-trial detection of gait initiation from movement related cortical potentials. Clin. Neurophysiol. (2014)

M. Jochumsen, I.K. Niazi, H. Rovsing, C. Rovsing, G.A.R. Nielsen, T.K. Andersen, N.P.T. Dong, M.E. Sorensen, N. Mrachacz-Kersting, N. Jiang, D. Farina, K. Dremstrup, Detection of movement intentions through a single channel of electroencephalography, 465–472 (2014)

A. Karni, G. Meyer, C. Rey-Hipolito, P. Jezzard, M.M. Adams, R. Turner, L.G. Ungerleider, The acquisition of skilled motor performance: fast and slow experience-driven changes in primary motor cortex. Proc. Natl. Acad. Sci. U. S. A. **95**, 861–868 (1998)

E.C. Leuthardt, G. Schalk, J. Roland, A. Rouse, D.W. Moran, Evolution of brain-computer interfaces: going beyond classic motor physiology. Neurosurg. Focus **27**, E4 (2009)

M. Li, Y. Liu, Y. Wu, S. Liu, J. Jia, L. Zhang, Neurophysiological substrates of stroke patients with motor imagery-based Brain-Computer Interface training. Int. J. Neurosci. **124**, 403–415 (2014)

J.M. McDowd, An overview of attention: behavior and brain. J. Neurol. Phys. Ther. **31**, 98–103 (2007)

M. Mukaino, T. Ono, K. Shindo, T. Fujiwara, T. Ota, A. Kimura, M. Liu, J. Ushiba, Efficacy of brain-computer interface-driven neuromuscular electrical stimulation for chronic paresis after stroke. J. Rehabil. Med. **46**, 378–382 (2014)

I.K. Niazi, N. Jiang, O. Tiberghien, J. Feldbæk-Nielsen, K. Dremstrup, D. Farina, Detection of movement intention from single-trial movement-related cortical potentials. J. Neural Eng. **8**, 066009 (2011)

I.K. Niazi, N. Mrachacz-Kersting, N. Jiang, K. Dremstrup, D. Farina, Peripheral electrical stimulation triggered by self-paced detection of motor intention enhances motor evoked potentials. IEEE Trans. Neural Syst. Rehabil. Eng. **20**, 595–604 (2012)

A. Pascual-Leone, J. Grafman, M. Hallett, Modulation of cortical motor output maps during development of implicit and explicit knowledge. Science **263**, 1287–1289 (1994)

A. Pascual-Leone, A. Amedi, F. Fregni, L.B. Merabet, The plastic human brain cortex. Annu. Rev. Neurosci. **28**, 377–401 (2005)

M.A. Perez, B.K.S. Lungholt, K. Nyborg, J.B. Nielsen, Motor skill training induces changes in the excitability of the leg cortical area in healthy humans. Exp. Brain Res. **159**, 197–205 (2004)

N. Petersen, H. Morita, J. Nielsen, Modulation of reciprocal inhibition between ankle extensors and flexors during walking in man. J. Physiol. (Lond). **520**, 605–619 (1999)

F. Pichiorri, F.F. De Vico, F. Cincotti, F. Babiloni, M. Molinari, S.C. Kleih, C. Neuper, A. Kubler, D. Mattia, Sensorimotor rhythm-based brain-computer interface training: the impact on motor cortical responsiveness. J. Neural Eng. **8**, 025020 (2011)

J. Polich, Updating P300: an integrative theory of P3a and P3b. Clin. Neurophysiol. **118**, 2128–2148 (2007)

A. Ramos-Murguialday, D. Broetz, M. Rea, L. Laer, O. Yilmaz, F.L. Brasil, G. Liberati, M.R. Curado, E. Garcia-Cossio, A. Vyziotis, W. Cho, M. Agostini, E. Soares, S. Soekadar, A. Caria, L.G. Cohen, N. Birbaumer, Brain-machine interface in chronic stroke rehabilitation: a controlled study. Ann. Neurol. **74**, 100–108 (2013)

J.J. Shih, D.J. Krusienski, J.R. Wolpaw, Brain-computer interfaces in medicine. Mayo Clin. Proc. **87**, 268–279 (2012)

S. Silvoni, A. Ramos-Murguialday, M. Cavinato, C. Volpato, G. Cisotto, A. Turolla, F. Piccione, N. Birbaumer, Brain-computer interface in stroke: a review of progress. Clin. EEG Neurosci. Off. J. EEG Clin. Neurosci. Soc. ENCS **42**, 245–252 (2011)

J. Song, B.M. Young, Z. Nigogosyan, L.M. Walton, V.A. Nair, S.W. Grogan, M.E. Tyler, D. Farrar-Edwards, K.E. Caldera, J.A. Sattin, J.C. Williams, V. Prabhakaran, Characterizing relationships of DTI, fMRI, and motor recovery in stroke rehabilitation utilizing brain-computer interface technology. Front. Neuroeng. **7**, 31 (2014)

D. Tome, F. Barbosa, K. Nowak, J. Marques-Teixeira, The development of the N1 and N2 components in auditory oddball paradigms: a systematic review with narrative analysis and suggested normative values. J. Neural Transm. **122**, 375–391 (2015)

E. Vaadia, N. Birbaumer, Grand challenges of brain computer interfaces in the years to come. Frontiers Neurosci. **3** (2009)

R. Xu, N. Jiang, C. Lin, N. Mrachacz-Kersting, K. Dremstrup, D. Farina, Enhanced low-latency detection of motor intention from EEG for closed-loop brain-computer interface applications. IEEE Trans. Biomed. Eng. **61**, 288–296 (2014)

R. Xu, N. Jiang, N. Mrachacz-Kersting, C. Lin, G. Asin, J. Moreno, J. Pons, K. Dremstrup, D. Farina, A closed-loop brain-computer interface triggering an active ankle-foot orthosis for inducing cortical neural plasticity. IEEE Trans. Biomed. Eng. (2014)

O. Yilmaz, W. Cho, C. Braun, N. Birbaumer, A. Ramos-Murguialday, Movement related cortical potentials in severe chronic stroke. Conf. Proc. IEEE Eng. Med. Biol. Soc. **2013**, 2216–2219 (2013)

B.M. Young, Z. Nigogosyan, L.M. Walton, J. Song, V.A. Nair, S.W. Grogan, M.E. Tyler, D.F. Edwards, K. Caldera, J.A. Sattin, J.C. Williams, V. Prabhakaran, Changes in functional brain organization and behavioral correlations after rehabilitative therapy using a brain-computer interface. Front. Neuroeng. **7**, 26 (2014)

K.A. Yurgil, E.J. Golob, Cortical potentials in an auditory oddball task reflect individual differences in working memory capacity. Psychophysiology **50**, 1263–1274 (2013)

U. Ziemann, T.V. Ilic, P. Jung, Long-term potentiation (LTP)-like plasticity and learning in human motor cortex–investigations with transcranial magnetic stimulation (TMS). Suppl. Clin. Neurophysiol. **59**, 19–25 (2006)

S. Aliakbaryhosseinabadi, L. Petrini, N. Mrachacz-Kersting, Effect of different attentional level and task repetition on movement-related cortical potential, September 16–19, 2014 at the Graz University of Technology, Austria (2014)

N. Mrachacz-Kersting, N. Jiang, R. Xu, K. Dremstrup, D. Farina, Plasticity following skilled learning and the implications for BCI performance, September 16–19, 2014 at the Graz University of Technology, Austria (2014)

Recent Advances in Brain-Computer Interface Research—A Summary of the BCI Award 2014 and BCI Research Trends

Christoph Guger, Brendan Allison and Gernot Müller-Putz

The preceding chapters summarized ten of the most promising BCI projects in 2014. As we said in the introduction, the jury had a difficult time selecting ten nominees—and three winners—out of the 69 submitted projects. 2014 was the first year we chose a second and third place winner, which only added to the competition and the excitement at our Gala Award Ceremony. The Award was presented at the 6th International Brain-Computer Interface Conference in Graz (Austria), and dozens of BCI researchers from around the world were present for the ceremony. Without further ado, we announce the three winners of the 2014 BCI-Research Award.

The 2014 Winners

The BCI Award 2014 winner is:

K. Hamada[a], H. Mori[b], H. Shinoda[a], T.M. Rutkowski[b,c] ([a]The University of Tokyo, JP, [b]Life Science Center of TARA, University of Tsukuba, JP, [c]RIKEN Brain Science Institute, JP)

Airborne ultrasonic tactile display BCI.

Gernot R. Müller-Putz, chair of the 2014 jury, called the winning idea "A fascinating new idea never explored before". This innovation could provide new BCI communication for persons without sight. This project, like other winners,

C. Guger (✉)
g.tec medical engineering GmbH, Guger Technologies OG, g.tec medical engineering Spain SL, g.tec neurotechnology USA, Inc., Schiedlberg, Austria
e-mail: guger@gtec.at

B. Allison
Cognitive Science Department, University of California, San Diego, CA, USA
e-mail: ballison@cogsci.ucsd.edu

G. Müller-Putz
Laboratory of Brain-Computer Interfaces, Graz University of Technology, Graz, Austria
e-mail: gernot.mueller@tugraz.at

© The Author(s) 2015
C. Guger et al. (eds.), *Brain-Computer Interface Research*,
SpringerBriefs in Electrical and Computer Engineering,
DOI 10.1007/978-3-319-25190-5_12

scored very high across several of the jury's selection criteria. The following figures show the first, second, and third place winners receiving their awards at the 2014 Gala Award Ceremony. In addition to certificates, the first place trophy and cash prize, winners also earned caps, shirts, and a copy of our book about the 2013 BCI-Research Award (Fig. 1).

The BCI Award 2014 2nd place winner is:

J. Ibáñez[a], J. I. Serrano[a], M.D. del Castillo[a], E. Monge[b], F. Molina[b], F.M. Rivas[b], J.L. Pons[a] ([a]Bioengineering Group of the Spanish National Research Council (CSIC), [b]LAMBECOM group, Health Sciences Faculty, Universidad Rey Juan Carlos, Alcorcón, ES)

Heterogeneous BCI-triggered functional electrical stimulation intervention for the upper-limb rehabilitation of stroke patients.

The 3rd place winner is:

N. Mrachacz-Kersting[a], N. Jiang[b], S. Aliakbaryhosseinabadi[a], R. Xu[b], L. Petrini[a], R. Lontis[a], M. Jochumsen[a], K. Dremstrup[a], D. Farina[b] ([a]Sensory-Motor Interaction, Department of Health Science and Technology, DK, [b]Dept. Neurorehabilitation Engineering Bernstein Center for Computational Neuroscience University Medical Center, DE)

The changing brain: bidirectional learning between algorithm and user (Fig. 2).

At the Gala Award Ceremony, Dr. Guger also thanked the experts in the 2014 jury:

Fig. 1 Christoph Guger (organizer), Brendan Allison (moderator), Peter Brunner (jury member), Gernot Müller-Putz (head of jury), Tomek Rutkowski (winner 2014), Katsuhiko Hamada (winner 2014)

Fig. 2 The person in the middle is Natalie Mrachacz-Kersting (3rd place 2014)

Gernot R. Müller-Putz (chair),
Deniz Erdogmus,
Peter Brunner,
Tomasz M. Rutkowski,
Mikhail A. Lebedev, and
Philip N. Sabes.

Directions and Trends Reflected in the Awards

The Annual BCI-Research Awards have helped to highlight the most promising trends in BCI research. By exploring the different properties of the submitted and nominated projects across different years, our book series has helped identify changes and new directions. The following four tables summarize different characteristics of submitted projects since the award began in 2010. In each table, N reflects the number of submissions, and numbers in different cells present the percentage or submissions with that characteristic. We present one table for each of the four general BCI components presented in the introduction.

Sensors: Table 1 explores the different types of input signals used in the submitted projects. As with previous years, the 2014 submissions focused primarily on EEG-based systems, similar to most BCI articles. The submissions also reflected other non-invasive sensor systems, such as fMRI and NIRS, and invasive methods like ECoG and neural spikes.

Table 1 Type of input signal for the BCI system

Property	2014 % (N = 69)	2013 % (N = 169)	2012 % (N = 68)	2011 % (N = 64)	2010 % (N = 57)
EEG	72.5	68.0	70.6	70.3	75.4
fMRI	2.9	4.1	1.5	3.1	3.5
ECoG	13.0	9.4	13.3	4.7	3.5
NIRS	1.4	3.0	1.5	4.7	1.8
Spikes	8.7	7.1	10.3	12.5	–
Other signals	4.3	13.0	2.9	1.6	–
Electrodes	–	6.5	1.5	1.6	–

Table 2 Real-time BCIs and off-line algorithms in projects submitted to the BCI Awards

Property	2014 % (N = 69)	2013 % (N = 169)	2012 % (N = 68)	2011 % (N = 64)	2010 % (N = 57)
Real-time BCI	87.0	92.3	94.1	95.3	65.2
Off-line algorithms	8.7	5.3	4.4	3.1	17.5

Signal processing: The second table analyzes the percentage of submissions that presented offline versus real-time BCI applications. Interestingly, 2014 reflected an ongoing recent trend toward offline BCIs. That is, some submissions described systems that did not operate in real-time, but could inspire real-time BCIs in the future. However, the substantial majority of submissions presented real-time tools (Table 2).

Output/application: The third essential component of any BCI is the output. Table 3 summarizes the different outputs, and related applications, that have been submitted since 2010. The applications have varied over the years, but generally show a strong interest in control, BCI platform tools, stroke and neural plasticity, and control of robotic devices such as prosthetics and wheelchairs. One emerging trend in 2014 was a substantial increase in submissions that addressed authentication and speech assessment. However, none of these submissions were nominated.

Environment/interaction: Finally, Table 4 summarizes the type of control signal that was used to influence BCI operation. This is a key component of the BCI's operating environment and interaction with each user. There were no major changes in 2014, with strong emphasis on evoked potentials (such as the P300 and SSVEP) and motor imagery. This is also consistent with general trends in BCI research. These three approaches have remained prevalent in BCI research since its early days.

Table 3 Type of output system and application

Property	2014 % (N = 69)	2013 % (N = 169)	2012 % (N = 68)	2011 % (N = 64)	2010 % (N = 57)
Control	17.4	20.1	20.6	34.4	17.5
Platform technology	13.0	16.6	16.2	9.4	12.3
Stroke neural plasticity	13.0	13.7	26.5	12.5	7
Wheelchair robot prosthetics	13.0	11.8	8.8	6.2	7
Spelling	8.7	8.3	25	12.5	19.3
Internet or VR game	2.9	5.9	2.9	3.1	8.8
Learning	5.8	5.3	1.5	3.1	–
Monitoring	1.4	4.7	4.4	1.6	–
Stimulation	1.4	3.6	1.5		
Authentification speech assessment	13.0	3	–	9.4	–
Connectivity	–	2.4	1.5	–	–
Music, art	1.4	1,8	–	–	–
Sensation	–	1.2	–	1.6	–
Vision	1.4	1.2	1.5	–	–
Epilepsy, parkinson, touret	2.9	1.2	–	–	–
Depression, fatigue	1.4	–	1.5	–	–
Neuromarketing, emotion	1.4	–	1.5	–	–
Ethics	1.4	–	–	–	–
Mechanical ventilation	–	–	–	1.6	–

Table 4 Type of control signal used to interact with the BCI

Property	2014 % (N = 69)	2013 % (N = 169)	2012 % (N = 68)	2011 % (N = 64)	2010 % (N = 57)
P300, N200	11.6	11.8	30.9	25	29.8
SSVEP/SSSEP/cVEP	11.6	14.2	16.2	12.5	8.9
Motor imagery	37.7	25.4	30.9	29.7	40.4
ASSR	–	1.8	–	1.6	–

Conclusion and Future Directions

BCIs are rapidly advancing. This book has presented research innovations to improve our understanding of brain function and help new groups of patients. Concordantly, the BCI research community has been growing, with new groups and publications from around the world. The outlook for BCI research is generally positive. However, many new directions still need additional work—and validation with patients—before they are ready to help broad groups of users. We expect that future research will highlight the most promising directions, and lead to new technologies to help people.

The Annual BCI-Research Awards have been successful in recognizing and encouraging high-quality BCI research innovations. We plan to continue the Award series, Gala Award Ceremonies, and book series. The 2015 BCI-Award was announced online, with a deadline of July 1, 2015. As of this writing, the jury is evaluating submissions. We are proud to announce the jury for 2015:

- Junichi Ushiba (chair of the jury 2015),
- Masayuki Hirata,
- Nuri Firat Ince,
- Zachary Freudenburg,
- José del R. Millán,
- Sydney Cash,
- Tomasz M. Rutkowski (winner 2014).

As with prior years, the first place winner from the preceding year is included among the top-notch jury. The jury again includes a strong international focus. For the first time, the chair comes from a top Japanese BCI institute. Dr. Ushiba has been an Assistant Professor at Keio University Faculty of Science and Technology since 2007, with a strong emphasis on BCIs for rehabilitation.

We expect that our introduction of second and third place winners will make future competition even more intense. Aside from cash prizes, nominees have additional opportunities to earn prestige and international recognition. Our 2015 flyer presents more information about the 2015 BCI-Research Award (Fig. 3).

We also plan to continue this book series to review and analyze the top BCI projects in 2015 and future years. We are pleased to announce that our editorial staff

Fig. 3 The flyer for the 2015 BCI-Research Award

for the book reviewing the 2015 BCI-Research Award will again include the CEO of G.TEC (Dr. Guger) and two other top BCI researchers: Drs. Allison and Ushiba. Like previous years, the third editor will be the chair of the jury for that year. We are looking forward to continuing our annual awards, ceremonies, and book series. This book series would not be practical without readers, and we hope you have enjoyed this book.